The Key to
Earth History

The angular unconformity described by James Hutton at Siccar Point, Berwickshire. Here Hutton first realised the antiquity of the Earth. (Drawing by Hilary Foxwell)

The Key to Earth History

An Introduction to Stratigraphy

PETER DOYLE, MATTHEW R. BENNETT

ALISTAIR N. BAXTER

School of Earth Sciences, University of Greenwich, UK

JOHN WILEY & SONS

Chichester • New York • Brisbane • Toronto • Singapore

Other Wiley Editorial Offices

John Wiley & Sons, Inc., 605 Third Avenue,
New York, NY 10158-0012, USA

Jacaranda Wiley Ltd, 33 Park Road, Milton,
Queensland 4064, Australia

John Wiley & Sons (Canada) Ltd, 22 Worcester Road,
Rexdale, Ontario M9W 1L1, Canada

John Wiley & Sons (SEA) Pte Ltd, 37 Jalan Pemimpin #05-04,
Block B, Union Industrial Building, Singapore 2057

Library of Congress Cataloging-in-Publication Data

Doyle, Peter.
    The key to earth history : an introduction to stratigraphy / Peter
    Doyle, Matthew R. Bennett, and Alistair N. Baxter.
        p.   cm.
    Includes bibliographical references and index.
    ISBN 0-471-94845-4
    1. Geology, Stratigraphic,   I. Bennett, Matthew (Matthew R.)
    II. Baxter, Alistair N.   III. Title.
    QE651.D64   1994
    551.7—dc20   94–325
                                                        CIP

British Library Cataloguing in Publication Data

A catalogue record for this book is available from the British Library

ISBN 0-471-94845-4

Typeset in 10/12pt Palatino from authors' disks by Production Technology Department,
John Wiley & Sons Ltd, Chichester
Printed and bound in Great Britain by Redwood Books Ltd, Trowbridge, Wiltshire

Contents

Preface

This book is a personal view of stratigraphy written by two lecturers in the field. Our view has developed through the experience of teaching a subject which is of fundamental importance and yet is traditionally viewed, by students at least, as boring; a simple recital of dates and rock units. We find this disappointing. In our book we attempt to present an account of stratigraphy as it really is: the key to understanding Earth history.

The book was written by Peter Doyle and Matthew Bennett during the summer of 1993 and is a direct result of many hours of both fruitful and, at times, fruitless discussion. The third member of our team, Alistair Baxter, provided a steadying hand and a guiding spirit. Alistair curbed our excesses of enthusiasm, provided guidance in many key areas and carefully read and edited the text throughout. Without his involvement this project would not have been completed.

We hope that by reading our book you will understand the process and pattern of Earth history. We also hope that our book will go some way to generate interest and enthusiasm for a subject which we feel is vital in understanding the development of our planet. If we can claim this, then we will have succeeded in our aim.

Peter Doyle, Matthew Bennett
and Alistair Baxter
London, 1993

Acknowledgements

The writing of any text inevitably draws upon the accumulated wisdom of innumerable colleagues. As a synthesis, our book is broad in approach, with little space for the explanation of detail. We apologise if this appears to neglect the valuable detailed work of many stratigraphers. We alone are responsible for the style, content and any errors or omissions.

There are many people to whom we are indebted for help, advice, encouragement and forbearance in producing this book. In particular, we would like to thank our colleagues at the School of Earth Sciences, University of Greenwich, for providing valuable comments on various drafts of this book, often at very short notice: in particular Andy Bussell, Paula Carey, Tony Grindrod and Geoff Manby. We thank our postgraduates, Tony Cosgrove and Jo Woodfin, for their help in many times of crisis. Two of our undergraduate students, Louise Broby and Chris Grimmer, provided a valuable student perspective on much of the text. A draft of the book was also reviewed externally by: Gillian Bennett, Paul Bown, Jane Francis, Trevor Greensmith, Duncan Hawley, Roger Mason and Colin Prosser. We wish to extend our thanks to these reviewers whose comments have done much to improve the clarity and accuracy of our message.

Nick Dobson and his library team at the University of Greenwich have dealt patiently with our often unreasonable demands, usually at short notice, for books, journals and other reference material. We are grateful for the technical support received from Hilary Foxwell, Martin Gay and Bill Ralph. We would also like to thank Amanda Hewes and Nicola Langridge at John Wiley who have guided this project from the start. Peter Doyle would like acknowledge the support and above else the patience of Julie who lived with the book as it was written.

Illustrations

The figures which accompany this book were drawn by Hilary Foxwell. Hilary has turned our crude roughs into superb line drawings and has developed a common style for the illustrations in line with our philosophy for the book.

Many of the illustrations are not original. Where they are reproduced exactly as they were first published, permission has been obtained from the copyright holder, and is indicated in the figure caption by the word *from* followed by the source. Every effort has been made to obtain permission from the relevant copyright holders, but where no reply was received we have assumed that there was no objection to our using the material. Where we have significantly altered an illustration, we have acknowledged this and signalled the change in the figure caption by using the phrase *modified from*.

We are grateful to the following for permission to reproduce copyright figures and photographs: The Natural History Museum (Figures 3.1, 3.6 and 4.9); Open University Press (Figure 5.1); Cambridge University Press (Figures 5.6 and 5.13); The Association of American Petroleum Geologists (Figure 5.16); Geological Society Publishing House (Figures 7.3 and 7.4); Chapman & Hall (Boxes 10.1 and 10.3); Merrill Publishing Company (Figure 10.4); the Palaeontological Association (Figure 10.6). Figures 3.2, 3.5, 3.7, 5.2A, 5.2C, 5.20A, 7.2 and 10.7 are reproduced by permission of the Director, British Geological Survey: NERC copyright reserved.

1
Introduction: What is Stratigraphy?

1.1 STRATIGRAPHY: THE KEY TO EARTH HISTORY

Geology is the science of the Earth. Its aim is to understand the composition, structure and history of the Earth throughout the 4600 million years of its existence. There are many different branches to geology: geochemistry, petrology and mineralogy are the branches concerned with the chemical and mineral composition of the Earth; sedimentology and geomorphology deal with the surface processes which shape the Earth's crust; geophysics looks at the structure of the Earth, both at the surface and at depth; palaeontology is concerned with the past life of the planet. These subjects have parallels with other physical sciences, but geology has the unique component of time: the development of the Earth in time and space. Within geology the study of time is the study of **stratigraphy**. Stratigraphy is the key to understanding the Earth's crust and its materials, structure and past life. It encompasses everything that has happened in the history of the planet. All geologists, whatever their speciality, are practitioners of stratigraphy: all geological study is an attempt to unravel piece by piece the history of Planet Earth.

Stratigraphy is also the art of detection. The rocks on the Earth's surface are the clues from which the Earth's history and the processes which have shaped it can be deduced. Each layer or **stratum** of rock contains a clue to the Earth's geography, climate and ecology at a specific time. The job of the stratigrapher is to **observe, describe** and **interpret** a sequence of such layers and other rock bodies in terms of events and processes in the history of the Earth.

1.2 THE AIM AND STRUCTURE OF THIS BOOK

Generations of geologists have come to believe that stratigraphy is boring: simply a recital of geological time periods and of different strata. Like human history, Earth history can be dull when taught as a series of facts and dates, but it does not have to be. When history is taught in terms of personalities and when both the cause and background to events are discussed, it can be fun, even exciting. The aim of this book is to illustrate that stratigraphy is dynamic and exciting. In order to achieve this the emphasis is placed here on understanding the patterns present within Earth history as opposed to the detailed succession of rock units which make up the stratigraphical record. In writing this book we have assumed a basic geological knowledge; however, some important principles are summarised at the end of this chapter (Boxes 1.1, 1.2, 1.3 and 1.4).

The book is split into two parts. In the **first part**, we look at the basic information needed to dissect the stratigraphical record: we explain the concept of 'uniformitarianism', present 'the stratigraphical tool kit' with which the stratigraphy of an area can be built up, discuss plate tectonics, and examine a series of case studies. In the **second part**, we set out the global pattern present within Earth history: a pattern which is driven by plate tectonics. The control of these global events on the stratigraphical record is then discussed and used to structure an account of the British Isles through geological time.

BOX 1.1: THE ROCK CYCLE

The rocks of the Earth's crust fall into three genetic groups: **igneous, metamorphic** and **sedimentary**. The creation of these three groups forms part of a natural cycle.

The primary rock types are **igneous rocks**. These form through the crystallisation of molten rock (**magma**) derived by melting of the Earth's interior. The texture of igneous rocks consists of interlocking crystals, which form as the rock cools and crystallises. The size of the crystals is dependent on the rate at which they have cooled: large crystals result from slow cooling while rapid crystallisation produces very small crystals.

When exposed at the surface, weathering and erosion break down these primary rocks into **sediments**, which become cemented through the processes of lithification into **sedimentary rocks**. Sedimentary rocks which are composed of the fragments of other rocks, are referred to as **clastic**, although some form through the precipitation of minerals from water (**evaporites**), or by the accumulation of shelly fossils (some **carbonates**).

Both igneous and sedimentary rocks may be subject to heat and/or pressure (**metamorphism**) after formation. This process creates **metamorphic rocks**, which have new minerals and textures which reveal the effects of pressure (e.g. the alignment of crystals) or temperature (e.g. interlocking crystal growth). Ultimately, all three rock types may be eroded to form sediments, or melted at depth to produce magma, completing the cycle.

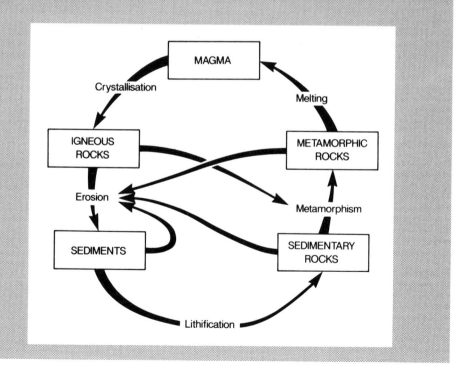

BOX 1.2: ROCK BODIES: FRACTURES AND FOLDS

The components of the Earth's crust consist of rock bodies which are either layered (**stratified**) or unlayered (**massive**).

Sedimentary rocks are usually stratified; igneous and metamorphic rocks, while sometimes appearing stratified, are more often massive crystalline bodies. Both types of rock body may show evidence of later deformation.

Rock bodies may be **folded**; this is where the traces of the layers are no longer horizontal, but curved. A fold is an expression of the plastic bending of a rock sequence, and may consist of an arched structure (known as an **anticline**) usually in combination with a reverse arch (known as a **syncline**). Folded sequences exhibit considerable complexity and provide a key, through the subject of **structural geology,** to understanding the architecture of the Earth's crust.

Rock bodies may also be deformed by **brittle deformation**; this is where the body is broken or fractured into a series of **faults** or **joints**. A fault is a fracture along which relative movement of the two parts has taken place, while a joint is a fracture along which no such movement has taken place. There are three basic types of fault, determined by the relative movement of their component parts: **normal, reverse** and **strike-slip**, as illustrated in the diagram below.

Both folds and faults are common in the stratigraphical record and help complete the picture of Earth development.

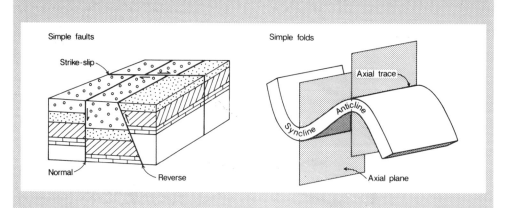

BOX 1.3: IGNEOUS ROCK BODIES

Igneous rocks are produced by the cooling of molten magma. Magma may either be **intruded** (forced into the Earth's crust), or **extruded** (forced out onto the Earth's surface). Igneous **intrusions** may take the form of large masses of igneous rock which are intruded deep in the crust (**plutons**, such as a **batholith**), or they may be smaller sheet-like bodies which exploit vertical fractures in the crust (**dykes**) or planes of weakness parallel to the stratified sequences into which they intrude (**sills**).

 Lavas are extruded onto the Earth's surface through **volcanoes** or through the exploitation of fractures in the Earth's crust.

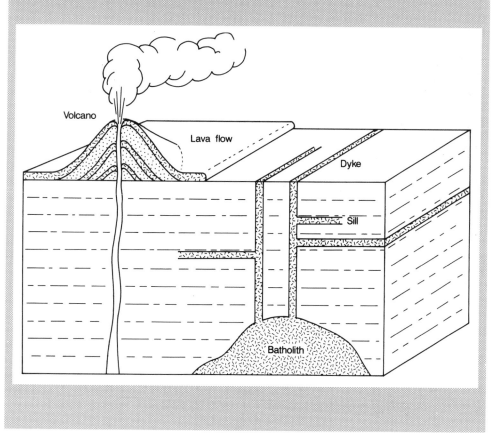

BOX 1.4: FOSSILS AND THEIR PRESERVATION

Fossils are the remains of once living animals and plants. Fossils are often referred to as **petrified** (i.e. turned to stone), but in fact the processes are a lot more subtle. The commonest fossils are those of animals or plants that lived in the sea and had hard sketetal parts or tough tissues. The rarest fossils are of creatures that lived on dry land, or were soft bodied, or had an easily fragmented skeleton. The **fossil record** is surprisingly complete, and in many exceptional examples, even the most delicate animals or plants have been preserved.

The process of fossilisation has three components:

- First, the animal or plant dies. It is relatively rare to discover the cause of death, but even this is possible in the right conditions.
- Second, the fossil must be buried. If it is not buried quickly enough, scavengers and biochemical processes will break down the flesh, and physical processes, such as wind and water, will lead to the breakdown of the skeleton.
- Third, once buried, the organism will undergo a variety of processes that will transform the organic material into mineral matter. Even then, given the right chemical environment, tissue and original skeletal material may be preserved unaltered. More commonly, shell and other skeletal material is impregnated or replaced by mineral compounds in solution in pore waters, or is dissolved completely and replaced wholesale by another mineral filling the cavity.

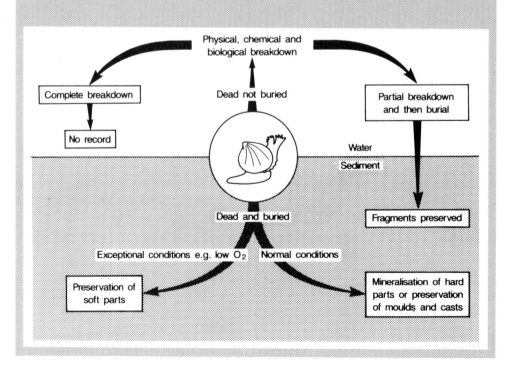

Part I
PRINCIPLES

2
Uniformitarianism or Actualism

In this chapter we introduce the principle of uniformitarianism and explain its importance as an expression of the scientific method adopted in the study of stratigraphy.

2.1 UNIFORMITARIANISM

Uniformitarianism is one of the most fundamental geological principles. It can be summarised in the phrase 'the present is the key to the past'. In the context of stratigraphy this means that the processes we observe today can be used to unlock the mysteries of Earth history.

James Hutton (1726–1797) was the first person to express this principle and grasp its significance for geological time. In 1785 he presented his *Theory of the Earth* in which he put forward a body of evidence to prove that the hills and mountains of today are being slowly eroded and that the sediment produced is being deposited as sand and mud on the sea floor to form sedimentary rocks. He argued that the processes of erosion and deposition which can be observed today have always operated. Hutton realised that the vast thickness of sedimentary rocks implied the operation of erosion and sedimentation throughout a period of time that he could only describe as being inconceivably long. This contribution was of immense importance at a time when the Earth was widely believed, on the basis of biblical estimates, to have formed at only 4004 BC.

In contrast, Georges Cuvier (1769–1832) interpreted the geological record in terms of a series of **catastrophes**. In 1796 he was the first person to recognise the presence of extinctions of fossil species within the geological record. To Cuvier, the constant replacement of species through time was clearly a function of major global changes or catastrophes. Cuvier recognised many of these catastrophes within the rock record, but did not invoke supernatural causes. William Buckland (1784–1856), however, interpreted Cuvier's findings in terms of a biblical framework, relating each event to a 'flood', the last of which was believed to be that experienced by Noah.

Charles Lyell (1797–1875) was not disposed towards invoking such catastrophic events in the explanation of Earth history. Lyell was strongly influenced by the ideas of Hutton and was convinced that the world had been shaped by gradual events and that rates of uplift, denudation and of deposition proceeded in a uniform pattern at a **constant rate**. The term which was coined for this view was 'uniformitarianism'. It was built in part upon Hutton's principle, and was formalised by Lyell in his *Principles of Geology* of 1830. However, Lyell assumed not only the **uniformity of process** or natural law as Hutton did but also the **uniformity of the rate** (gradualism). Lyell therefore excluded temporary and local crises, though Hutton's original doctrine did not. During the middle of the 19th century there was a continuum between, on the one hand, the extreme catastrophists who invoked both present-day and unknowable or supernatural causes and, on the other, the extreme gradualists, such as Lyell, who allowed only present-day processes operating with their present intensity. Lyell's ideas eventually triumphed and dominated the theoretical development of stratigraphy to such an extent that any theory which allowed for the existence of catastrophic influences was deemed unscientific.

In recent years, however, there has been a growing awareness that the stratigraphical record appears to contain not only sediments produced by slow gradual processes but also those deposited by more catastrophic events. This does not mean that the Earth was formed through a series of catastrophies, but reflects the greater probability that catastrophic events are preserved in the stratigraphical record. This can perhaps be best explained in terms of two concepts taken from the study of surface processes. The first of these two concepts is the interrelationship of magnitude and frequency: the greater the magnitude of an event the less frequent its occurrence. Small events may occur several times each year while large events may only occur once in every 100 or 1000 years. For example, hurricanes do not happen every year in southern Britain, but they do occur occasionally, say once every 100 years or so. If you examine the meteorological records of a single year it is therefore unlikely that a hurricane will be recorded, but if you were to examine the records for a period of over 1000 years then the chances of a hurricane being recorded are much greater. In a geological context the longer the time period under consideration, the more likely that the record will contain 'high magnitude' or catastrophic events. We should not, therefore, base our judgement on human time scales. The second relevant concept is that, in general, the bigger the event, the greater its effect and therefore the longer it will take to remove the record left by that event, and the more likely it is that it will be preserved in the sedimentary record. This point is well illustrated in the peat bogs along the east coast of Scotland which contain a record of a catastrophic wave or tsunami; a magnitude of event which appears to have only occurred once in Britain's recent Quaternary record (Box 2.1). Smaller magnitude storm events are not recorded in the sedimentary sequence, but this catastrophic event is.

The result of these observations has, however, led to 'gradualism' being nowadays largely rejected in a geological context. Uniformitarianism expressed in terms of the unity of process—or **actualism** as it is called in continental Europe—is, however, still the fundamental concept which underlies any attempt to provide a scientific explanation of Earth history. In this context, actualism can be summarised in the statement

BOX 2.1: A TSUNAMI DEPOSIT IN EASTERN SCOTLAND: AN EXAMPLE OF CATASTROPHISM

Within the sequence of Quaternary sediments along the east coast of Scotland there is a thin, regionally extensive sand layer which has been dated as 7000 years old. This sand layer is to be found in the peat bogs and clays (estuarine muds) of the Scottish coast. It extends locally more than two kilometres inland from the former coastline and some four metres above the former sea level. This sand horizon is believed to have been deposited by a tidal wave centred on the North Sea. The tsunami was probably caused by a small earthquake which triggered a submarine landslide, known as the Storegga slide, in the northern part of the North Sea. It is this landslide which is believed to have produced the tidal wave deposit of the east coast of Scotland some 7000 years ago. The movement of large volumes of sediment and the subsequent sudden changes in sea bed topography can cause dramatic fluctuations in sea level. If the entire volume of the slide failed instantaneously as a block, the theoretical maximum resulting wave may have had a height of 350 m. The actual wave was probably of much lower height since the failure does not appear to have been instantaneous. This catastrophic event is well recorded in the sedimentary record because the sand layer has been deposited well inland of the area in which conventional marine processes could have eroded it. This catastrophic event is recorded in the sedimentary record, but more conventional events such as storms and tidal surges are not. It clearly illustrates the way in which the sedimentary record may be biased towards catastrophic events and may not therefore adequately reflect the ordinary geomorphological or depositional processes operating.

Source: Long, D., Smith, D. E. & Dawson, A. G. 1989. A Holocene tsunami deposit in eastern Scotland. *Journal of Quaternary Science* **4**, 61–66.

that 'no powers are to be employed that are not natural to the globe, no action is to be admitted except those which we understand and can observe'.

Given that the aim of stratigraphy is to reconstruct the changing geography, climate and events of the Earth through time, there are three main tasks to be tackled: (1) establishing the chronological sequence of rocks; (2) establishing a time history for these rocks; and (3) interpreting them in terms of the environment in which they were deposited, that is, in terms of their palaeogeography, palaeoclimate and palaeoecology. Each of these three tasks is dealt with in the following chapters.

2.2 SUMMARY OF KEY POINTS

• In the early 19th century the geological record was widely interpreted in terms of a series of **catastrophic** events, of which many were explained in terms of supernatural causes or within a biblical framework.

3
Establishing the Sequence
of Events

Stratigraphy is about interpreting the Earth's geological record as a sequence of events through time. As each sequence of rocks is equivalent to a sequence of events, it is necessary to have the tools to interpret and understand the relative order of these events. In this chapter we explain the two basic principles of relative ordering, **superposition** and **relative chronology**, and then show how they can be applied to establish the sequence in the field.

3.1 THE PRINCIPLE OF SUPERPOSITION

The **principle of superposition** states that in any undisturbed sedimentary sequence, the layer at the bottom of that sequence was formed first. Superposition is therefore the most fundamental principle in determining the relative order of rock units in a sequence.

The principle was developed in 17th century, when Niels Stensen (1638–1686), a Danish physician living in Florence, was invited to dissect a large shark. He was careful to observe and describe the shark in minute detail. In the course of his work, Stensen, often known by his anglicised Latin name of Steno, recognised that the teeth of the shark closely resembled the stony fossils, known as Glossopetrae, found in the uplands of Tuscany. At this time the idea that fossils were the preserved remnants of past organisms was not widely held, but Steno by careful observation demonstrated that the Glossopetrae and the shark's teeth were identical in form (Figure 3.1).

LAMIAE PISCIS CAPVT·

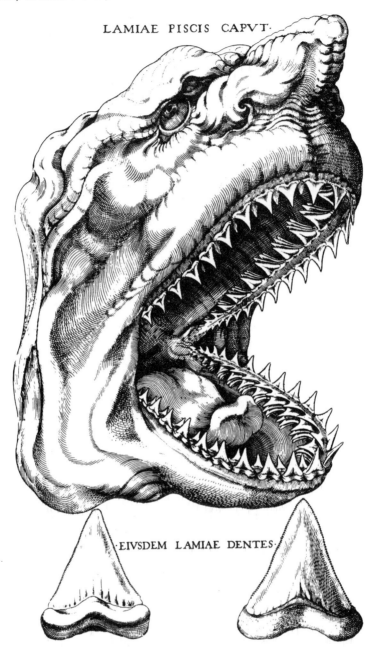

·EIVSDEM LAMIAE DENTES·

Figure 3.1 *Steno's comparison of shark's teeth with Glossopetrae. The living shark was dissected by Steno and the teeth compared with the fossil objects known as Glossopetrae, illustrated beneath the shark's head. This illustrated clearly for the first time the organic nature of fossils.* [From: Mercati (1717) Metallotheca, The Vatican, p. 333. Reproduced with the permission of the Natural History Museum, London]

This discovery demonstrated that fossils were actually the remains of once-living animals, and that the rocks of Tuscany were formed in a sea which had once covered the land. Steno recognised that fossils were actually incorporated in rocks before they were lithified, and that the deposition of sediments in the sea ultimately led to the formation of fossil-bearing rocks in layers. The outcome of Steno's investigations was the development of the principle of superposition.

Steno published the results of his work in 1669 and his results can be summarised in three principles:

1. **Superposition:** when a layer of rock was forming, the one below had already formed and was therefore older.
2. **Original horizontality:** layers of rock were originally deposited horizontally.
3. **Original lateral continuity:** layers of rocks extend laterally until physically constrained in some way.

Using these three principles, Steno was able to propose a **relative chronology** for the layers of strata within the Tuscany area. A chronology is a list of events in the order in which they occurred.

3.2 SUPERPOSITION

The principle of superposition allows us to identify the order in any sequence of undisturbed sedimentary rocks which are stacked in layers one upon another; the oldest will always be found at the bottom (Figure 3.2). Determining the chronology of rock units in a vertical sequence involves establishing the relative time relationships between each unit of the sequence. The discovery of superposition provides the mechanism by which bedded rocks can be recognised as the products of a chronological sequence of depositional events. Today, most geologists take this principle for granted, but its importance cannot be underestimated. From it, geologists have a simple method by which they can determine the relative order of undisturbed strata. The Grand Canyon in Arizona provides a good example of the principle of superposition in action (Figure 3.3). Here an unusually complete geological section through the Earth's sedimentary rock record can be seen in the sides of the canyon cut by the Colorado River. At the base there are ancient rocks which underlie a sequence of sedimentary layers, stacked one on top of another. Application of the principle of superposition leads us to conclude that the oldest layer of rock is at the base and that each successive layer decreases in age. The sequence of rock layers can be summarised as a stratigraphical column (Figure 3.3).

Even in deformed sequences, where the layers have been folded and the definition of the original top and bottom of the sequence is ambiguous, superposition can be applied with the aid of what are known as **right way-up structures**. These structures provide a record of the original way-up of the sedimentary sequence by defining the original top and bottom of the sequence. Examples of right way-up structures are given in Figures 3.4 and 3.5, while their application in the field is illustrated in Box 3.1.

Figure 3.2 *Sequence of horizontally bedded strata (Blue Lias) from the Lower Jurassic of Lyme Regis, Dorset, England. [Photograph: BGS]*

3.3 DETERMINING RELATIVE CHRONOLOGY

On its own the principle of superposition cannot resolve the chronology of complex geological areas, in which the rocks are, for example, not simply in horizontal layers. In such areas geologists require more sophisticated tools with which to establish the relative chronology of both rock units and geological events. The most important of these tools are **cross-cutting relationships** and **unconformities**. Before looking, however, at each of these tools in detail we shall first look at the development of these ideas by early geologists.

Superposition as a tool was applied literally by many early geologists. Geologists such as Abraham Gottlob Werner (1749–1817) argued that geological strata continued infinitely around the globe and were precipitated from a primeval ocean which once covered the globe. Each layer was horizontal and indicative of Steno's principle of superposition. Werner divided the world's geological succession into four units: (1) the **Primitive** or crystalline rocks which were precipitated from this primeval and universal ocean; (2) the **Transition** rocks containing fossils indicating the creation of life; (3) the **Flötz** deposited in running water formed by the contraction of this universal ocean; and (4) the **Alluvial** rocks formed by the flow of water on land. Werner

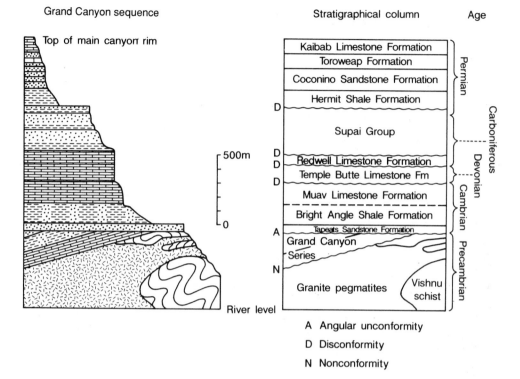

Figure 3.3 *Stratigraphy of the Grand Canyon sequence. The Grand Canyon exposes a series of horizontally bedded strata overlying much older, folded and metamorphosed rocks. Application of the principle of superposition allows us to construct a stratigraphical column. [Modified from: Press & Siever (1986) Earth, Freeman, Fig. 2.14, p. 32]*

interpreted all rocks, including crystalline granites and basalts, as being formed by sedimentation in an ocean or by running water. Implicit in his model of the geological record was the fact that deformation of rock layers and surface erosion had only occurred once within Earth history (during the Flötz and Alluvial Periods). This model of Earth history was consistent with the ideas of the theologians of the time who considered the Earth to have been formed just six thousand years ago.

The first person to really appreciate that Werner's scheme was inadequate to explain the development of the Earth was James Hutton. Hutton approached the interpretation of the geological record on the basis of sound field observations backed by deductive reasoning. His researches lead him to question the ideas of Werner and his contemporaries. He made two important observations.

First, at Glen Tilt, in the Cairngorm Mountains of Scotland, Hutton noted that granite appeared to penetrate the surrounding metamorphic schists in a series of veins (Figure 3.6). He concluded that the granite must have been in a molten state for it to have forced its way into the schist in this way. Hutton made two inferences from this conclusion: (1) granite is not precipitated from water as suggested by Werner and his

Examples of way-up criteria

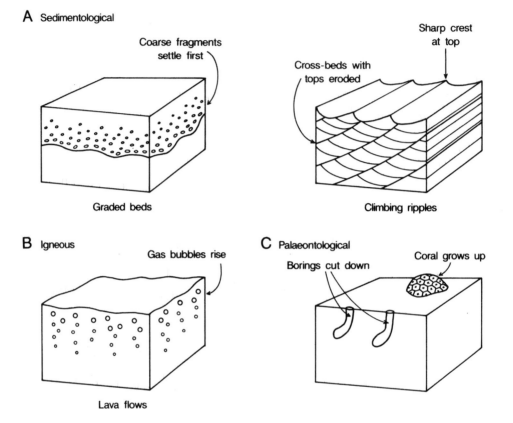

A Sedimentological

Coarse fragments
settle first

Cross-beds with
tops eroded

Sharp crest
at top

Graded beds

Climbing ripples

B Igneous

Gas bubbles rise

Lava flows

C Palaeontological

Borings cut down

Coral grows up

Figure 3.4 *Examples of way-up criteria.* **A:** *Sedimentological—graded beds and ripples. When sediment settles out of suspension from a body of water, the coarser particles settle more quickly than the finer ones which produces a characteristic type of bedding known as graded bedding (coarse at the bottom, fine at the top). Inverted sequences would show coarser particles at the top not the bottom. Current or tidal ripples on the surface of sediments may be preserved, and the right way-up is indicated by the sharp ripple crests which always point upwards. The sedimentary structures produced by ripples can also be used to indicate way-up, because the tops of the cross-beds produced by ripples are often eroded (see Fig. 5.2).* **B:** *Igneous—in a lava flow, gas bubbles rise to the surface. Therefore, the concentration of bubbles in a lava flow illustrates the original way-up.* **C:** *Palaeontological—corals and similar organisms grow upwards towards the light, while burrows are cut downwards into the sediment*

contemporaries, but is formed by cooling of molten rock and (2) that the granite must be younger than the schists into which it was forced as a fluid mass. The second of these inferences is an example of a cross-cutting relationship and is important in determining the relative chronology of geological events as we shall see in the next section.

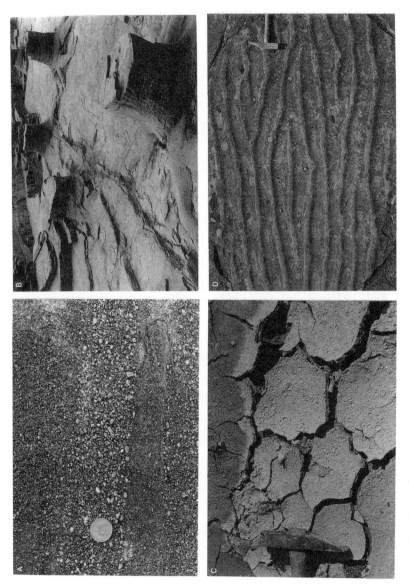

Figure 3.5 *Photographs of typical way-up structures.* **A:** *Graded bedding in the Torridonian Sandstone of north-west Scotland. Within each unit of graded bedding the particles are coarse at the bottom and become finer upwards [Photograph: BGS].* **B:** *Fossil trees of Lower Carboniferous age, near Glasgow (Lepidodendron trees—Lycopods). Trees grow upwards; therefore, in situ fossil trees provide good way-up evidence [Photograph: BGS].* **C:** *Recent sedimentary features formed on a muddy puddle in the floor of a quarry by the action of rain (rain pits) and sun (desiccation cracks). Both these structure may be preserved in the sedimentary record and can be used as way-up criteria [Photograph: BGS].* **D:** *Ripple marks on a bedding surface of Weald Clay Sandstone, from West Sussex, England. The sharp crests of the ripples point upwards [Photograph: BGS]*

BOX 3.1: WAY-UP CRITERIA IN STRATIGRAPHICAL STUDIES

In 1958 R. M. Shackleton was first to demonstrate that the rocks in the Aberfoyle anticline were actually upside down. This anticline forms part of an area of the Highland borders in Scotland that was intensely deformed during the Grampian Orogeny. Although the rock sequences involved have been heavily affected by the tectonic activity, the sequence of turbidite sandstones known as the Aberfoyle Grits still retain their primary sedimentary structures and, particularly, their graded bedding. Shackleton demonstrated that this graded bedding was consistently reversed in Aberfoyle, and from this he concluded that the structure was actually inverted and formed part of the much larger Loch Tay belt—a large, complexly overfolded structure associated with the Grampian Orogeny. This example demonstrates the value of detailed stratigraphical studies to the understanding of the development of a complex part of the Earth's crust.

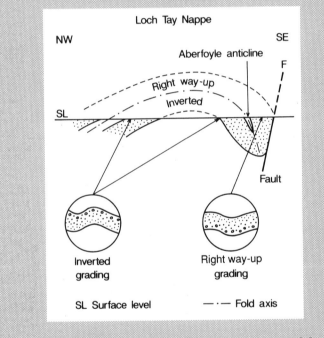

Source: Shackleton, R. M. 1958. Downward-facing structures of the Highland border. *Quarterly Journal of the Geological Society of London* **138**, 361–392.

Second, at Siccar Point in Berwickshire, southern Scotland, and at other localities on the coasts of Scotland and northern England, Hutton made a discovery that illustrated that the age of the Earth was much in excess of that estimated on the basis of the biblical account of creation and that periods of rock deformation, uplift and erosion had occurred more than once in the geological record. At Siccar Point, Hutton observed two bodies of strata apparently at right angles to each other. The older shale

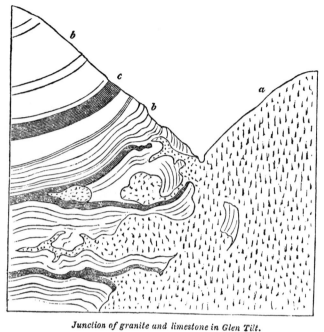

Junction of granite and limestone in Glen Tilt.

a, Granite.   *b,* Limestone.   *c,* Blue argillaceous schist.

Figure 3.6 *Lyell's (1830) illustration of Hutton's famous locality at Glen Tilt, Cairngorm Mountains, showing the true nature of granite. Hutton appreciated from this site that the granite must have been intruded in a molten state into the schists and limestones of the surrounding area. This cross-cutting relationship clearly illustrates the relative age of the granite, which is younger than the surrounding schist rocks. [From: Lyell (1830) Principles of Geology, John Murray, Fig. 88. Reproduced by permission of the Natural History Museum, London]*

strata were arranged so that the beds were almost vertical, while the younger overlying sandstone strata lay almost horizontally on top of the vertical beds (Figure 3.7). In view of the fact that the two units did not conform to the normal pattern of one horizontal layer on top of another the relationship was termed an **unconformity**. Importantly, fragments of the older strata were found in the lower bed of the topmost formation. If sedimentary rocks were laid down originally on a horizontal plane, as Steno had taught, then the underlying sequence must have undergone a cycle of deposition, tilting and erosion, before the overlying rocks, which incorporated blocks of the older strata, were then deposited on top. As John Playfair (1742–1819), one of Hutton's contemporaries, put it in 1802, 'what clearer evidence could we have had of the different formation of these rocks and of the long interval that separated their formation'. Hutton drew two inferences from these observations: (1) that there had been several periods of deformation, uplift and erosion, not one as suggested by Werner and (2) the Earth must be of much greater antiquity than the predictions made from the Bible to allow sufficent time for this to occur.

Figure 3.7 *Photograph of Hutton's unconformity at Siccar Point, Berwickshire, southern Scotland. Unconformity of Upper Old Red Sandstone (Devonian) on the vertical Silurian greywackes and shales. [Photograph: BGS]*

Cross-cutting relationships and **unconformities** are therefore very important in determining the relative chronology of geological strata and are discussed further in the following sections.

3.3.1 Cross-cutting Relationships

Hutton's discovery at Glen Tilt established the principle that rock sequences are older than the geological features or structures, such as faults or igneous intrusions, which cut across them. The **principle of cross-cutting relationships** is particularly important in determining the chronology of rock units where other independent data are unavailable. Figure 3.8 shows a typical example of a geological sequence in which cross-cutting relationships are present. In this example the sedimentary sequence of layers 1 to 6 were deposited first. The igneous body, known as a dyke (D), was intruded after the sedimentary succession was deposited. This chronology is apparent because of the way that the igneous body cuts across the sedimentary succession. Within larger igneous bodies, fragments of the surrounding sedimentary succession may be incorporated into the molten rock as it is forced into position. These fragments, known as

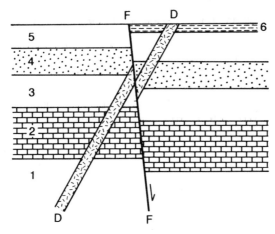

Figure 3.8 *An example of cross-cutting field relationships; see text for explanation*

xenoliths, help confirm that the surrounding succession, the **country rocks**, were formed first. Continuing with our example, a fault (F) subsequently developed and this cut and displaced not only the original sedimentary layers but also the later igneous dyke (Figure 3.8). Therefore the principle of cross-cutting relationships allows us to recognise three events in our example (Figure 3.8) with the following relative chronology: (1) deposition of beds 1 to 6; (2) intrusion of the dyke; and (3) formation of the fault and displacement along it.

The principle of cross-cutting relationships is extremely valuable as a methodological tool for establishing relative chronology and is particularly useful in rock successions which have been subsequently folded, faulted and intruded by igneous rock. A detailed example of the application of this tool in the geological record is presented in Section 7.1.1 of this book.

3.3.2 Unconformities

Unconformities allow the recognition of a relative chronology of events, but also indicate the presence of time gaps within rock sequences. These gaps may span very long or very short intervals of time. Four types of unconformity can be recognised in the stratigraphical record, each formed under different circumstances. The four types are (Figure 3.9): (1) **Angular unconformity**, (2) **Disconformity**, (3) **Nonconformity**, and (4) **Paraconformity**.

Hutton's original unconformity at Siccar Point (Figure 3.7) displays a marked angular discordance between the rock layers beneath and above the plane of unconformity, and provides therefore a good example of an **angular unconformity**. Put crudely, the rock units above and below the plane of unconformity have different angles of dip. The history of this unconformity involves the following sequence of processes: deposition→deformation→uplift→erosion→redeposition. This cycle of

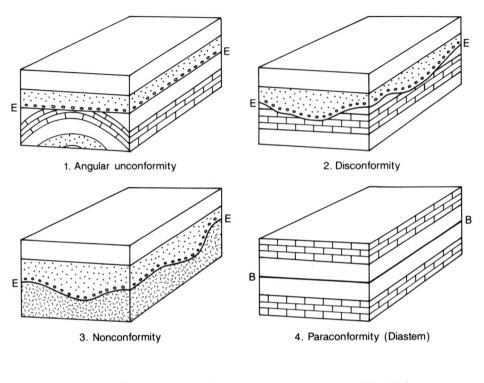

1. Angular unconformity   2. Disconformity

3. Nonconformity   4. Paraconformity (Diastem)

E ⌇ Erosional surface   B ── Depositional break   ∘∘∘∘∘ Eroded fragments

Figure 3.9 *The four main types of unconformity. [Modified from: Dunbar & Rodgers (1957), Principles of Stratigraphy, John Wiley & Sons, Fig. 57, p. 117]*

events is illustrated in Figure 3.10 and can be broken down into four stages. First, a sequence of sedimentary rocks was deposited horizontally, according to Steno's principle of superposition. Second, this sequence was then folded and uplifted, probably as a result of mountain building caused by continental collision (see Section 6.5). Third, the uplifted and folded rocks were subject to erosion and the landscape was dissected and lowered. Finally, a rise in sea level flooded this eroded landscape and a new sequence of sedimentary rocks was deposited horizontally above the older deformed sequence.

The three other types of unconformity are more subtle. A **disconformity** occurs where the units above and below the plane of unconformity have the same angle of dip, but are separated by a surface which has undergone erosion (Figure 3.9). Here, the older sequence of rocks has not been folded, but has been subject to erosion before the deposition of the younger rocks above the erosion surface. This type of unconformity may occur because of a fall in sea level which exposes the rocks of a sea floor to subaerial erosion, before a subsequent sea level rise leads to the deposition of new sedimentary rocks on top of the erosional surface. The erosion surface represents a

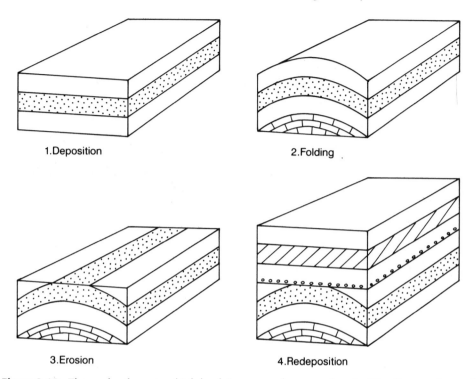

1.Deposition

2.Folding

3.Erosion

4.Redeposition

Figure 3.10   *The cycle of events which leads to an angular unconformity. The diagram shows the four phases in the production of an angular unconformity: (1) the deposition of horizontally bedded sediments; (2) folding and uplift of the sedimentary strata; (3) the erosion of the folded strata; (4) the deposition of new sedimentary rocks on top of the eroded surface*

break or time gap in sedimentation. A classic example of a disconformity is seen in the Hampshire and London Basins of southern England. Here an eroded and potholed Cretaceous Chalk surface denotes a major erosional break in the sequence between the Cretaceous Chalk and the younger Eocene sediments above (Figure 3.11). (See Figure 4.8, page 52 for the divisions of geological time.)

Where bedded sediments overlie an exposed and eroded crystalline basement, the term **nonconformity** is used (Figure 3.9). Often this is indicative of a landscape formed by the erosion of intensely deformed and metamorphosed rocks produced during continental collision, although it can be produced when a large igneous intrusion is exposed by erosion. Deposition of horizontally layered sedimentary rocks, during a rise in sea level, or by river systems, produced the nonconformity. In Britain a good example of this is the deposition of the Torridonian red sandstones of the Late Proterozoic on top of the much older, dissected ancient land surface of the crystalline Lewisian basement, in north-west Scotland.

Linking all of the unconformities discussed so far is the fact that each is characterised by evidence of erosion during which time there was no deposition within the area. Consequently the erosional surface represents a time break within the sedimentary

Figure 3.11 *Photograph of the disconformity at the base of the Palaeogene deposits of the London Basin. The photograph illustrates the uneven surface of the eroded Cretaceous Chalk strata at Pegwell Bay, Kent, England. This surface is overlain by younger, bedded sands, deposited during the Palaeogene [Photograph: D. G. Helm]*

sequence. In many cases, fragments may be eroded out of the underlying sequence and re-incorporated or **reworked** into the covering rocks. However, we can also recognise an unconformity, or time break, where there has been little or no erosion, but where the net rate of sedimentation has matched the net rate of sediment loss and consequently there is no sedimentary record of the time period in question. This type of unconformity is known as a **paraconformity**, or is sometimes called a **diastem**, where the gap is a small one. In these cases the sedimentary sequence is not exposed to subaerial erosion.

The gaps produced in the sedimentary record by non-deposition can be of many different orders of magnitude, representing a few hours, years or even millions of years. The length of time represented by such gaps can only be determined by recourse to independent means of dating. Some geologists believe that most of the stratigraphical sequence contains numerous diastems. In fact it has been argued that the stratigraphical record is rather like a net—a series of holes held together by string—and that each bedding plane is in fact a period of time when there was no deposition and is therefore a gap in the geological record. Attempts have been made to quantify the completeness of the stratigraphical record, with varying success (Box 3.2). It is important to realise that the record is imperfect, even in deep ocean basins where there are few bottom currents to cause erosion and where sedimentation is usually continuous.

BOX 3.2: ESTIMATING STRATIGRAPHICAL COMPLETENESS

Most geologists accept that stratigraphical sections are not a complete record of sedimentation through time. In fact, it is widely held that there are 'more gaps than record'. In making detailed interpretations it is necessary to have a clear appreciation of the size and extent of the gaps present, or alternatively to know the completeness of a given stratigraphical sequence. Therefore, to aid in detailed studies, some way of quantitatively assessing the completeness of the stratigraphical record is needed. Completeness can be defined as the proportion of the total time span being considered during which deposition took place. A yardstick of completeness can be created by estimating the rate of accumulation in a given environment, say for instance the annual sediment load of a major river, and calculating the volume of sediment that would accumulate in a specific basin over a specific time interval. Estimates of the completeness of the stratigraphical record for a given time span can then be calculated through the comparison of the observed thickness of a stratigraphical section with that expected from the accumulation rate and volume. In practice this would entail the comparison of a number of stratigraphical sections, taking the thickest as the most complete. McShea and Raup (1986), in doing this, were surprised to find that the thickest sections may in fact be representative of the time elapsed, implying that there are few gaps within the record. However, critics such as Tipper (1987) point out that in practice this technique is probably difficult to apply as it depends on being able to obtain reliable estimates of the rate of sediment accumulation, which is difficult.

Sources: McShea, D. W. & Raup, D. M. 1986. Completeness of the geological record. *Journal of Geology* **94**, 569–574. Tipper, J. C. 1987. Estimating stratigraphic completeness. *Journal of Geology* **95**, 710–715.

3.4 ESTABLISHING THE SEQUENCE IN THE FIELD

The first task of any geologist entering a new field area is to **establish the sequence of rocks** present. This involves three things: (1) detailed observation and description of the rocks (lithologies) present; (2) formal establishment of stratigraphical units according to convention; and (3) the determination of their distribution in space. This process is the subject of **lithostratigraphy**.

Lithostratigraphy is the formal description of rock units and their comparision with others in both space and time. It provides much of the raw data needed to interpret the geological record. It is fundamental to the process of comparing the stratigraphical record of one area with another in order to build up a picture of a region's geology. It is also the first stage in determining the time relationship of rocks in one area with those of another distant area.

Lithostratigraphy entails the division of rock sequences into units, known as **lithostratigraphical units,** on the basis of lithology (rock type). Each of these units should be lithologically homogeneous and their recognition is helped when they have distinct lithological boundaries. Traditionally, lithostratigraphical units have been defined for only layered or stratified rocks. The geological record is, however, made up of more than just sedimentary rocks and lithostratigraphy should encompass the description of all rock units, including igneous and metamorphic rocks. The term **lithodemic unit** is used by some authors when referring to rock units which are not stratified. This term has not, however, found universal favour.

Lithostratigraphy employs a strict hierarchy of terms for units so that geologists worldwide understand what is meant by each term (Table 3.1). This terminology is governed by international agreement. The fundamental unit is the **formation,** which is described as a unit of largely homogeneous lithology that may be clearly recorded on a geological map. A formation is therefore a **mappable unit.** It is only established formally when it has been described and published in the geological literature. This ensures that different geologists use the same name for the same unit, which prevents confusion and misinterpretation. This formal procedure is described in Box 3.3. Description of these mappable units forms the basis for **lithostratigraphy.**

Table 3.1 *The hierarchy of formal lithostratigraphical units*

Lithostratigraphical unit	Definition
Supergroup	Clustering of groups based on their lithological characteristics or on their mode of formation
Group	Clustering of formations based on their lithological characteristics or on their mode of formation
Formation	Mappable unit of homogeneous lithology
Member	Subdivision of a formation
Bed	Lithologically distinct horizon or layer

Lithostratigraphical units are said to **crop out** at the Earth's surface and it is usual to refer to their lateral extent at the surface as their **outcrop.** The distribution of lithostratigraphical units is best recorded on an **outcrop map** (i.e. a **geological map**). The construction of such maps is an important part of the stratigrapher's task and one which is made difficult by the fact that the outcrop of a geological unit is often covered by soils, vegetation or buildings. Geological maps are therefore usually made by study of, and extrapolation from, the rocks which are actually visible, or **exposed,** at the surface.

The comparison of lithostratigraphical units with others in different locations is known as **correlation.** Usually comparison is made with sequences in adjacent areas, but correlation may be achieved between widely spaced localities, for example between continents. The ultimate aims of correlation are: (1) to establish the relative

BOX 3.3: FORMAL LITHOSTRATIGRAPHY IN PRACTICE

The Lower Jurassic rocks of North Yorkshire, in northern England, are excellently exposed in a series of coastal exposures. Consequently, they have been the focus of much attention since the birth of geology. Earlier geologists referred to these rocks as the Lias, a name possibly derived from an ancient quarryman's term for layered rocks. The lithostratigraphy of these rocks developed through the *ad hoc* addition of informal units. In recent years it has become apparent that this informal lithostratigraphy was inadequate on two counts. First, many of the traditional units were not clearly defined with respect to lithology and, second, it was not clear which stratigraphical sections were the most representative of each unit. In view of this, Powell (1984) proposed to formalise the lithostratigraphy of these rocks according to modern convention. He recognised five formations, instead of the previously unsatisfactory Lower, Middle and Upper divisions of the Lias. Each formation is a **mappable unit**, with an internally homogeneous lithology and distinct boundaries, and each can be mapped at the surface as well as being easily recognisable in boreholes by geophysical methods. Each unit was formally defined with reference to five parameters: (1) the **lithology** was described; (2) **type** or **reference sections** were indicated with map coordinates to facilitate future study; (3) **formal names** were chosen on the basis of a place, a lithology and a unit rank (e.g. the **Whitby Mudstone Formation**) to properly characterise each unit; (4) the nature of mappable **boundaries** was discussed; and (5) the **thickness** of each unit was recorded. This is illustrated in the table. Powell (1984) therefore provides an excellent example of how a formal lithostratigraphy should be established.

	Name	Lithology	Type section	Boundaries	Thickness
Lias Group	Blea Wyke Sandstone Formation	Sandstones	Blea Wyke Point	Sandstone	
	Whitby Mudstone Formation	Grey–dark grey mudstones	Whitby	Grey mudstone / Grey mudstone	*c*.105m
	Cleveland Ironstone Formation	Mudstones with ironstone seams	Staithes	Iron-rich mudstone / Mudstone/ironstone	*c*.28m
	Staithes Sandstone Formation	Fine–medium sandstones	Staithes Harbour	Sandstone / Fine-grained sst	*c*.21m
	Redcar Mudstone Formation	Grey mudstones	Robin Hood's Bay	Mudstone	*c*.250m

Source: Powell, J. H. 1984. Lithostratigraphical nomenclature of the Lias Group in the Yorkshire Basin. *Proceedings of the Yorkshire Geological Society* **45**, 51–57.

chronology of lithostratigraphical units and therefore the relative sequence of geological events over a wide area and (2) to provide an understanding of the geological development of a specific area through a knowledge of the spatial pattern of rock units within it. In practical terms such correlation may be achieved by a variety of means. Simple visual matching of lithological succession is the commonest method used. **Lithological equivalency** is established by drawing **tie lines** (Figure 3.12) to link lithological units which have a similar composition. Tie lines make no assumption of time equivalency; the same rock type may form in two different areas at different times (Figure 3.12). In contrast, **time lines** are independent of lithology and are established using time-significant tools such as **fossils** or sudden geological **events**; these tools are discussed in Chapter 4. Tie lines may therefore cross time lines.

Sometimes lithological correlation may be achieved by **geophysical methods**. Such methods involve correlation of rock units on the basis of their electrical or other physical properties. These methods are particularly important, for example, where correlation is to be achieved between a series of borehole cores (Figure 3.13). These methods provide a signal which is indicative of the physical properties of the rock (e.g. **electrical resistivity**). Correlation is achieved by drawing tie lines between matching signals (Figure 3.13).

In order to understand the distribution, form and significance of lithological units it is important to appreciate three important concepts: (1) **distribution in space**, (2) **distribution in time** and (3) **relative thickness**.

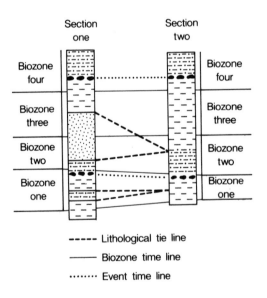

Figure 3.12 *Physical correlation and the relationship of tie lines and time lines. The sedimentary sequences in sections one and two can be correlated on the basis of lithological similarity. The lithological units are linked with tie lines. Time lines link points of equal age. They may be constructed on the basis of faunal evidence (biozones), in which case they are independent of lithology and time lines may cross tie lines. Time lines may also be based on lithological event horizons, or isochronous surfaces, in which case they will not cross tie lines*

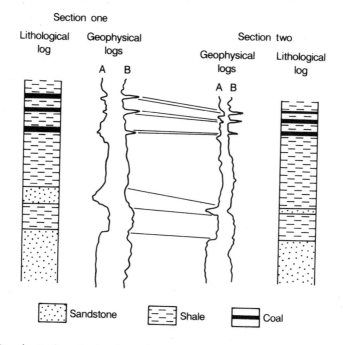

Figure 3.13 *Geophysical methods of correlation. The diagram illustrates how the geophysical properties, which are dependent on lithology, can be used to correlate sedimentary sequences. The two lines A and B illustrate varying electrical properties of the different lithological units. Tie lines are drawn between points with similar geophysical properties*

The **distribution in space** or **continuity** of a single unit will depend on the continuity of the original sedimentary environment in which it was deposited. For example, the sediments deposited in marine environments are often much more uniform in their lateral extent than those deposited in a fluviatile (river) environment, the boundaries of which are determined by those of the river valley. This is discussed further in Section 5.1.

Horizontal layers of sedimentary rocks may accumulate in one of two ways: (1) by vertical accumulation, due to a gentle rain of sediments from above, or (2) by lateral accumulation from specific points, such as the mouth of a river, from which sediments build out. Most contemporary sediments, away from deep ocean basins, accumulate laterally. Within the geological record sedimentary layers which accumulated laterally may be indistinguishable from those which accumulated vertically. This has implications for the **distribution in time** of lithological units. Units which have accumulated laterally in space will have formed at progressively different times in different places. Lithostratigraphical units which accumulated in this way are not therefore equivalent to time units and are said to be **diachronous**—that is, although the unit is uniform in lithology, it has formed at different times in different places. Another problem exists when considering the distribution of lithostratigraphical units in time. Given the range of depositional environments present today, it follows that we would expect a similar range of environments, with similar products, to have recurred through geological time. For example, desert sandstones have formed more than once in Earth history and their occurrence is not therefore time significant. As a consequence we should not assume that because two rock units are of the same lithological type, they were formed at the same time. In other words, lithostratigraphical units are not usually in themselves time significant; in order to compare units of the same age an independent means of correlation is needed (Figure 3.12). The most common tools for correlation are fossils, and their use as stratigraphical tools is known as **biostratigraphy**, a technique which is discussed in Section 4.1.2.

The **relative thickness** of a deposit or unit may also be misleading. There are many instances of immensely thick deposits in one area having been deposited over the same time interval as deposits of only a few metres thickness elsewhere. For example, the Lower Jurassic Junction Bed, exposed on the south coast of England in Dorset, is only one metre thick but is the time equivalent of the Whitby Mudstone Formation which is over 50 metres thick and exposed in north east England. This difference reflects deposition in two different types of sedimentary basin—a subsiding marine basin in the case of the Whitby Mudstone Formation (Yorkshire) and a submerged topographically high area in the case of the Junction Bed (Dorset). Thin successions like the Junction Bed, which have much thicker time-equivalent deposits elsewhere, are said to be **condensed** (Box 3.4). In summary, no assumption of time should be made from the thickness of a unit.

3.5 SUMMARY OF KEY POINTS

- The **principle of superposition**, which states that in an undisturbed sequence the lowest sedimentary layer is the oldest, allows the first recognition of the relative stratigraphical order of a series of rock units. The application of **way-up criteria** ensures the order of superposition can be accurately determined even in deformed and inverted sequences.
- Determining **relative chronology** enables a more detailed breakdown of stratigraphical sequences. **Cross-cutting relationships** are particularly important in metamorphic and igneous terrains. **Unconformities** help to establish sequences of events in geological time, and record the presence of time gaps.
- The first task in stratigraphy is the recognition, description and correlation of lithological units. This is known as **lithostratigraphy**.
- Many lithological units accumulated laterally and are therefore fundamentally **diachronous**: they are lithologically, but not temporally, identical throughout their extent.
- The relative thickness of correlated units is not necessarily indicative of the time taken in deposition. Immensely thick units may be deposited rapidly: relatively thin units may have accumulated slowly.

BOX 3.4: BASINS AND SWELLS IN THE JURASSIC

The Lower and Middle Jurassic rocks of England were deposited over a broad area of flooded continental shelf, and comprise a range of mudrocks and carbonates. Although the topography of this flooded area is thought to have been low, there are some parts which experienced greater amounts of subsidence than others. The areas experiencing the least amount of subsidence, known as **swells**, seem to equate with regions that were either buoyed-up by igneous intrusions (e.g. Market Weighton) or had underlying structural anomalies. The observed result in these positive areas is that stratigraphical sequences are relatively **condensed** in comparison with the adjacent subsiding basins and their **expanded** sequences. The accompanying figure illustrates the concept. The thickness of individual biozones denoted by fossil ranges provides a way of directly comparing the thickness of sequences in specific areas. In the Market Weighton area (locality 2) and the Mendip Hills (locality 4), for example, the topography was relatively high and the Jurassic rocks are condensed, being largely thin or absent. In contrast, in the adjacent Yorkshire (locality 1) and Cotswold (locality 3) basins there is a greater thickness of sediment preserved, representing an expanded sequence.

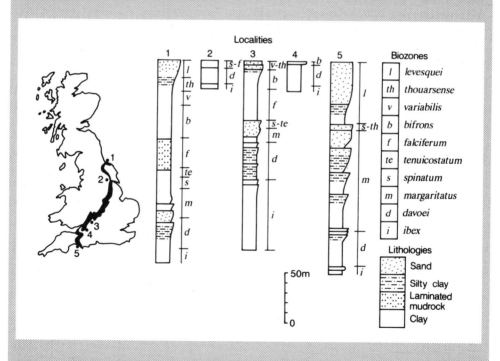

Source: Sellwood, B. W. & Jenkyns, H. C. 1975. Basins and swells and the evolution of an epeiric sea (Pliensbachian–Bajocian of Great Britain). *Journal of the Geological Society of London* **131**, 373–388. [Diagram modified from: Sellwood & Jenkyns (1975) *Journal of the Geological Society of London* **131**, Fig. 1, p. 374]

• Correlation between rock units may be achieved through matching lithologies, or through comparing the geophysical properties of individual lithologies. The lines of correlation joining lithologies are known as **tie lines.** Tie lines do not imply time equivalency, which can only be determined using independent, time-specific tools, such as fossils or geological events.

3.6 SUGGESTED READING

Ager (1993) is a thought-provoking and highly influential reader which should form the basis for further discussion of the nature of the stratigraphical record. Dott & Batten (1988) and Duff (1993) give attention to the basis of stratigraphical sequences. Boggs (1987) deals in detail with the concepts explored in this chapter, but can be complex in places. Whittaker *et al.* (1991) deal with the nomenclature and definition of stratigraphical units, while Barnes (1988) gives a detailed guide to the description of rock units in the field.

Ager, D. V. 1993. *The Nature of the Stratigraphical Record.* (Third Edition) John Wiley & Sons, Chichester.

Barnes, J. 1988. *Basic Geological Mapping.* Geological Society of London Handbook, Open University Press, Milton Keynes.

Boggs, S. 1987. *Principles of Sedimentology and Stratigraphy.* Merrill, New York.

Dott, R. H. & Batten, R. L. 1988. *Evolution of the Earth.* McGraw-Hill, New York.

Duff, P. McL. D. (Ed.) 1993. *Holmes' Principles of Physical Geology.* (Fourth Edition) Chapman & Hall, London.

Whittaker, A., Cope, J. C. W., Cowie, J. W., Gibbons, W., Hailwood, M. R., House, M. R., Jenkins, D. G., Rawson, P. F., Rushton, A. W. A., Smith, D. G., Thomas, A. T. & Wimbledon, W. A. 1991. A guide to stratigraphical procedure. *Journal of the Geological Society of London* **148**, 813–824.

4
Geological Time

We have seen in the previous chapter how we can interpret the relative chronology of rock units through the principle of superposition and by using cross-cutting relationships and unconformities. These tools are appropriate for limited stratigraphical studies, but in order to determine the time equivalence of units across the globe, we require an independent means of comparing rock units in relative time. These methods, such as **faunal succession** and **geological events**, are discussed in this chapter. They have allowed geologists to construct a standard stratigraphy, known as the **Chronostratigraphical Scale**, which ensures geologists can construct an accurate picture of global Earth history. This scale does not rely upon the recognition of absolute dates (i.e. in years), but on the relative order of geological units. Absolute dates may, however, be determined for stratigraphical units using techniques of **radiometric dating**, which are discussed at the end of this chapter.

For most geologists geological time is a relative concept and involves simply placing geological units in their relative order, and it is not therefore possible to talk in terms of precise ages in years. Absolute dates can be obtained but are not part of the day-to-day routine of most geological study.

4.1 FAUNAL AND FLORAL SUCCESSION

Life has existed on the Earth for at least 3500 million years, during which time it has evolved and developed irreversibly. Different species of animals (**fauna**) and plants (**flora**) have appeared, evolved and become extinct through time. The evidence for this succession of species is recorded in the fossils within the rock record. This principle is known as the **faunal and floral succession** and can be used to provide relative dates for geological units, since evolution is time dependent and irreversible. We shall first consider how the early geologists developed their understanding of the fossil record and applied it in the interpretation of geological successions.

James Hutton was able to determine crude relative chronologies and therefore interpret geological sequences within small areas, but was unable to build up a broad regional picture of the development of the geological record because he lacked the tools with which to **correlate**, or match, sequences of rock in one area with those in another. An independent means of correlation was needed.

This was to be provided by an English engineer William Smith (1769–1839). In 1793 he began work on the construction of a canal near Bath, in southern England. Smith built up a first-hand knowledge of the strata surrounding Bath and its order of superposition. He later set out to construct the first detailed geological map of England and Wales, which was published in 1815. Smith's work was based on personal field observation and was unencumbered with unnecessary theoretical considerations. The geological units chosen by Smith for his map were based on two criteria: (1) that the lithologies or rock types were distinctive and (2) that they contained a distinctive assemblage of fossils. The significance of Smith's observations was that, for the first time, strata were identified by their distinctive fossils and that this was used as a tool in tracing the map units over great distances.

In 1796, Georges Cuvier, working at the Paris Museum of Natural History, presented a paper *On the Species of Living and Fossil Elephants* to a meeting of eminent scientists in Paris. This was a significant paper because it was the first to suggest that species of animals had become extinct in the geological past. In this paper Cuvier compared the skeletons of recently discovered fossil elephants from Siberia and northern Europe with the skeletons of living Indian and African elephants. He concluded that the mammoth did not belong to the same species as either of the living elephants and, since there were no living representatives, it must be extinct. Cuvier was able to show with later research that this phenomenon was not isolated and that many fossils represented species that were long since extinct. Cuvier favoured catastrophic causes for the extinctions that he recognised, but it was not until the publication of Charles Darwin's *Origin of Species* in 1859 that evolution was presented as the all-encompassing theoretical basis for the successions of different fossils recognised by Smith, Cuvier and others.

Smith's discovery that strata could be identified by the fossils contained in them together with Cuvier's discovery of species extinction paved the way for the realisation that fossils could be used as indicators of relative age irrespective of sedimentary rock type. The first application of this concept was in the study of the relatively young rocks of the Cenozoic exposed in continental Europe. In the 1830s Deshayes in France, Bronn in Germany and Lyell in England all produced subdivisions of the Cenozoic that were based on fossils. In this way the foundations for the use of fossils as an independent means of correlation, and of biostratigraphy, were laid.

4.1.1 The Tools of Biostratigraphy

Biostratigraphy is the branch of stratigraphy which involves the study and interpretation of fossils to enable the correlation of sedimentary rock units.

The tools of biostratigraphy are fossils known as **guide fossils** (also **index** or **zone fossils**). The most useful guide fossils are those which, when living, were widely distributed both geographically and environmentally and which therefore allow a variety of rocks formed in widely spaced localities and in different environments to be correlated. As few fossil species fulfil all these ideal requirements, it follows that not all fossils are of equal value in biostratigraphy.

It is possible to draw up a set of criteria that can be used as a kind of 'check-list' to the suitability of fossils as correlation tools. Guide fossils should ideally be: (1) **independent of their environment**, (2) **fast evolving**, (3) **geographically widespread**, (4) **abundant**, (5) **readily preserved** and (6) **easily recognisable**. Few fossils qualify for honours in all of the criteria but it follows that the more they possess the better they are as guide fossils (Figures 4.1 and 4.2).

Fossils should be **independent of their environment** to be of value in inter-regional correlation. Bottom-dwelling marine animals, for example, may be dependent on the nature of the sea bottom sediment (**substrate**) or water depth. Fossils which are strongly dependent on a narrow range of environmental parameters are of limited use as guide fossils, because they only occur in these specific environments. For example, barnacles are restricted to rocky foreshores, while corals are strongly restricted by light, salinity and temperature, and both are therefore of limited value as guide fossils (Figure 4.1).

Rates of evolution are known to vary through geological time. Some fossils evolved quickly (**fast-evolving**) with the development of new species (**species turnover**) at regular intervals of less than one million years. High rates of species turnover, that is of evolutionary change, mean that a given fossil will exist within a smaller proportion of the stratigraphical record. For example, if a species survived for a long time then it could be found in rocks of widely different ages, but if it survived for only a short span of time its presence in a rock ties that rock to a more specific period of time. Consequently organisms with rapid species turnover allow the geological record to be subdivided more finely than ones in which new species only appeared after long intervals. The so-called **living fossils**—such as *Nautilus* (the pearly nautilus), *Lingula* (an intertidal brachiopod) and *Limulus* (the king crab)—have remained unchanged in overall morphology for vast periods of time and are therefore of very limited value in biostratigraphy.

Fossils with restricted environmental niches are known as **facies fossils**. Despite their inadequacy some facies fossils have been pressed into service as guide fossils. Corals in particular have been used where a uniform environment is achieved over a wide area, such as in the Lower Carboniferous rocks of England. In contrast, organisms that are independent of substrate, such as those adapted to a free-swimming or floating mode of life are well-suited for correlation because they may float or swim above, and therefore drop into, a variety of different depositional environments. Such organisms are the best guide fossils (Figures 4.1 and 4.2).

The **geographical extent** of fossils obviously controls the area over which correlations can be made. Thus if an organism is worldwide in its distribution, then there is the potential for worldwide correlation of stratigraphical sequences.

The last three criteria are important because they control whether a fossil can be easily employed as a guide fossil, irrespective of whether it is naturally suited to the

Criteria / Fossil	Independent of environment	Fast to evolve	Geographically widespread	Abundant	Readily preserved	Easily recognised	Status as guide fossils
 Graptolites	✔ (Plankton)	✔	✔ (Plankton)	✔	✔	✔ (Simple form)	Good (Ordovician to Silurian)
 Ammonites	✔ (Free swimming)	✔	✔ (Free swimming)	✔	✔	✔ (Great diversity)	Good (Devonian to Cretaceous)
 Corals	X (Need warm shallow sea)	X	X	✔	✔	✔	Poor (Carboniferous)
 Echinoids	X (Bottom dwelling)	X	X	✔	✔	✔	Poor (Cretaceous)
 Barnacles	X (Need rocky shore)	X	X	X	X	✔	Bad (not used)
Foraminifera	✔ (Plankton)	✔	✔ (Plankton)	✔	✔	✔	Good (Particularly Mesozoic to Recent)
Pollen	✔ (Wind blown)	✔	✔ (Wind blown)	✔	✔	✔	Good (Cretaceous to Recent)
 Coccoliths	✔ (Plankton)	✔	✔ (Plankton)	✔	✔	✔	Good (Mesozoic to Recent)
Birds 	✔ (Flying)	X	✔ (Flying)	X	X (Fragile bones)	✔	Bad (not used)

Figure 4.1 *Examples of good and bad guide fossils. The matrix illustrates how different fossil groups match up to the ideal criteria for a good guide fossil. It is important to note that each criterion is not necessarily of equal importance. For example, preservation potential can be of greater importance than widespread distribution. Bird fossils, otherwise well-suited as guide fossils, are rarely preserved and therefore, make bad guide fossils*

Figure 4.2 *Photographs of a selection of guide fossils.* **A:** *Ammonite* (Lytoceras) *from the Upper Jurassic of Antarctica—actual diameter 75 mm [Photograph: P. Doyle].* **B:** *Belemnite* (Gonioteuthis) *from the Upper Cretaceous of southern England—actual length 75 mm [Photograph: P. Doyle].* **C:** *Graptolite* (Didymograptus) *from the Ordovician of South Wales—actual length 35 mm [Photograph: R. Fortey].* **D:** *Foraminiferan* (Neogloboquadrina) *from the Holocene of the South Atlantic—actual diameter 225 μm [Photograph: F. L. Lowry].* **E:** *Trilobite* (Calymene) *from the Silurian of England—actual length 49 mm [Photograph: R. Fortey].* **F:** *Coccosphere* (Coccolithus), composed of individual coccoliths from the present-day North Atlantic—each coccolith is 10 μm in diameter [Photograph: J. Young]*

task. However, fossils should also be **abundant**: they will be of little use if they satisfy the first three criteria, but are impossible to find. The first bird Archaeopteryx, for instance, is otherwise well-suited as a Jurassic guide fossil, but only eight specimens are known to exist. Similarly, if fossils are too delicate, or their morphology too bland, or too complex, then practising stratigraphers will be unable to use them effectively. In general birds have hollow bones which are easily broken; although otherwise well-suited as guide fossils, they are limited by their poor fossil record (Figure 4.1). Graptolites preserve a simple morphology which can be recognised in even deformed rock sequences, and are excellent guide fossils (Figures 4.1 and 4.2). Fossils must therefore be **readily preservable** and **readily recognisable** to be of value.

Amongst the best fossils which can be employed in biostratigraphical correlation are ammonites—an extinct group of molluscs that died out at the end of the Meso-zoic—and graptolites—an extinct group of colonial organisms common in the Palaeo-zoic (Figures 4.1 and 4.2). Ammonites were free-swimming organisms that had a widespread distribution and which were not tied to a particular substrate. Importantly, ammonites evolved rapidly with the production of new species and the extinction of old ones each within an estimated time span of 0.5 to 1 million years. The practical result of this for biostratigraphy is that refined, high-resolution correlation can be achieved which allows fine subdivision of the Mesozoic stratigraphical record. Grap-tolites were free-floating organisms which have been used for intercontinental corre-lation of Palaeozoic sequences (Box 4.1). Graptolites evolved quickly and, like ammonites, were not restricted to a substrate or particular type of marine environment. Both groups were abundant, readily preserved and are easily recognisable with the aid of standard reference works.

BOX 4.1: LAPWORTH'S GRAPTOLITE BIOZONES

In 1878 Charles Lapworth erected a series of biozones for the Lower Palaeozoic sediments (greywackes) of the Southern Upland hills in Scotland. These biozones were based upon graptolites, animals whose true affinities were poorly understood at that time. Lapworth used graptolites as stratigraphical ciphers which helped to unlock the story of the development of the Southern Uplands, despite the relative complexity of their geological structure, deformed by plate tectonic processes. These ciphers proved to be of immense importance and were applied widely in interpre-ting other sequences of similar age. Lapworth's biozones have stood the test of time and of stratigraphical fashion. Graptolites are now much better understood as animals. Their planktonic lifestyle means that they are found in both deep and shallow water settings, they are also exceptionally widespread in distribution and they display rapid evolutionary development. As a result of this graptolites are ideal guide fossils. Their value is illustrated by the durability of Lapworth's original biozone scheme, which dates from the late 19th century, and is still used today.

Source: Fortey, R. 1993. Charles Lapworth and the biostratigraphic paradigm. *Journal of the Geological Society of London* **150**, 209–218.

Microfossils are also particularly important in biostratigraphy, particularly as they are small enough to be extracted from the small-diameter rock-cores produced from boreholes. This is obviously an advantage in the interpretation of underground sequences drilled in the quest for oil or other natural resources. Particularly important guide microfossils include foraminifera (single-celled shelly organisms) and pollen (Figures 4.1 and 4.2).

4.1.2 The Prime Unit of Biostratigraphy: The Biozone

The prime unit of biostratigraphy is the **biozone**, often just referred to as a **zone** in older literature. Biozones are strata organised into stratigraphical units on the basis of their content of guide fossils. Biozones may be recognised on local or regional scales.

The concept of the biozone was developed by Albert Oppel (1831–1865) in the 1850s. Oppel recognised that the vertical (stratigraphical) range of fossils was time significant and that it was independent of the lithology containing the fossils. He subdivided the Jurassic System into units defined on the vertical ranges of a number of fossil assemblages. In many ways this has parallels with Smith's original work, but Oppel differed in not tying the fossil assemblages to lithology. Each of Oppel's zones was formally defined and named, usually after one distinctive fossil in an assemblage, and each zone could be traced across continental Europe.

Oppel's concept of the zone, now re-christened the biozone to avoid ambiguity, remains valid today and some types of assemblage biozones are given the name Oppel biozones. In an **assemblage biozone**, the biozone is defined on the basis of the vertical ranges of a number of fossils (Figure 4.3). Often assemblage biozones are utilised where there are few good guide fossils available as, for example, is often found where most of the fossils have a benthonic (bottom-dwelling) rather than planktonic (free-floating) or nektonic (free-swimming) mode of life. An assemblage biozone is often limited in use because it relies on the recognition of a number of fossils. Other types of biozone have also been recognised. **Total range biozones** are based on the total vertical range of a single fossil, usually a suitably qualified guide fossil, such as an ammonite or graptolite. **Partial range biozones** utilise part of the total vertical range of a fossil, particularly of organisms which were relatively slow to evolve. The partial range is usually defined as being between the first and last appearances of other fossils (Figure 4.3). Other biozones, such as the **acme biozone**, which is based on an abundance of a fossil group, are more difficult to recognise in practice because of the overall imperfection of the fossil record.

The lower boundaries of biozones are usually drawn at the first appearance of the guide fossil, and the upper at the first appearance of the next one (Box 4.2). These boundaries are sometimes related to an **evolutionary lineage** where the replacement of species is directly related to the evolution of one species to another. Such biozones are called **consecutive range biozones** (Figure 4.3). In practice, they are hard to prove and rely upon detailed stratigraphical study. In other cases the boundaries of biozones may be denoted by migrations, which mark the migration of a species from one geographical area to another and may be unrelated to the organisms of the preceding

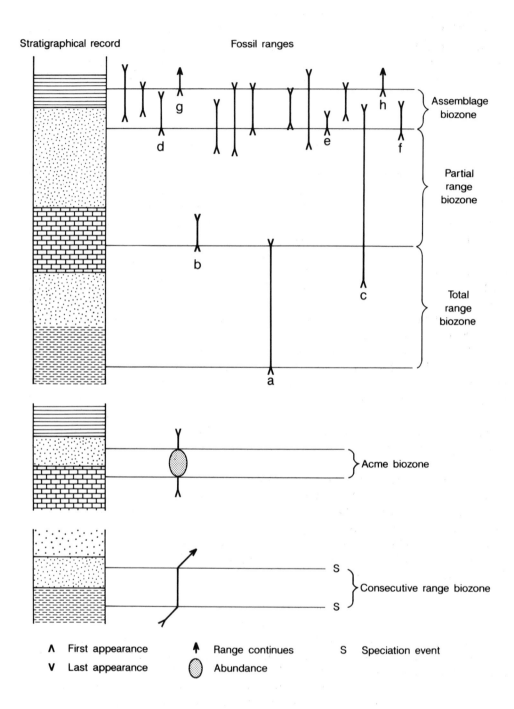

biozone. In practice, this is relatively common. The theoretical basis of any **biozonation scheme** is that the biological changes recorded in the scheme should be synchronous across the world. A new species will not evolve and appear everywhere in an instant but will migrate progressively from a centre across a region or continent. Consequently the appearance of a new species may not be synchronous across an area. However, this is not a problem given the sheer scale of geological time and that the process of species migrations will not be detectable within the resolving power of the geological record. They can, therefore, be considered for most purposes to be instantaneous.

In terms of evolution, the concept of **punctuated equilibria**, whereby species undergo periods of rapid change between long periods of stasis, has been recognised from the stratigraphical record (Box 4.3). Many zonal schemes may in fact be a reflection of such dramatic evolutionary change, although very detailed stratigraphical study is required to prove it.

4.1.3 Lithostratigraphy and Biostratigraphy Compared: The Problem of Diachronism

As we have seen in Chapter 3, the distribution of lithostratigraphical units in time and space is determined by the nature of the depositional environment. Although the relationship of one environment to another is often one of equilibrium—such as the relationship between beach, intertidal sand and offshore mud in a typical shallow marine setting—the relative positions of these environments may change through time. The deposits produced, for example during a sea level rise or fall, or during the lateral movement of a delta, will result in a lithostratigraphical unit which is internally homogeneous, but which has been formed at different times over its lateral range. In these cases, the resulting deposits transgress time boundaries and are said to be **diachronous**. Recognition of diachronism requires clear understanding of the faunal or floral succession which established the time lines. In this way the tie lines of correlation of diachronous units are seen to cross the time lines established by independent evidence such as guide fossils (Figure 3.12).

Diachronism is probably extremely common in the stratigraphical record, but often cannot be detected because the resolving power of biostratigraphy is insufficient. In fact, as we have seen in Chapter 3, in most cases, lithological units often develop laterally, accumulating from specific points to form a progressively continuous unit or layer. If each unit has developed laterally on a geologically insignificant scale, then it

---

Figure 4.3 *Types of biozone. The diagram shows the following types of biozone: (1) Consecutive range biozone, which is defined on the basis of the consecutive ranges of fossils in an evolving lineage. (2) Acme biozone, which is defined on the total abundance of an individual species. (3) Total range biozone, which is defined by the total vertical range of species (a). (4) Partial range biozone with the lower boundary defined by the first appearance of species (b), and the upper boundary by the first appearance of species (d) and (e). (5) Assemblage biozone, which is defined on the basis of the combined ranges of a number of fossils: [Modified from: Holland et al. (1978) A Guide to Stratigraphical Procedure, Geological Society of London, Special Report No. 10, Fig. 2, p. 14]*

BOX 4.2: SHAW'S METHOD OF GRAPHICAL CORRELATION

In 1964, the American stratigrapher A. B. Shaw devised a new method of semi-quantitative correlation of biostratigraphical sequences. Shaw's method utilises a single stratigraphical section which is taken as the standard. This standard is selected because it is the thickest and most unaffected by structural complications. From this section the fossil species ranges are constructed, with particular importance being assigned to the first appearance of fossil species, but also with reference to the last appearance. A graph can be constructed using the same data collected from another stratigraphical section. The first and last appearances (bottom and top of their stratigraphical range) of the fossil taxa provide points on the graph, and a best-fit line is constructed. If a best-fit line can be constructed, the sections can be correlated (Diagram A). The technique thus provides a clear graphical method which aids in the correlation of sequences. Changes in the gradient of the graph (dog-legs) indicate changes in the rate of sedimentation (Diagram B), while gaps in the record are illustrated by plateaux (Diagram C). This method is also applicable to event stratigraphy, and both data sets can be plotted on the same curve to provide independent corroboration of correlation.

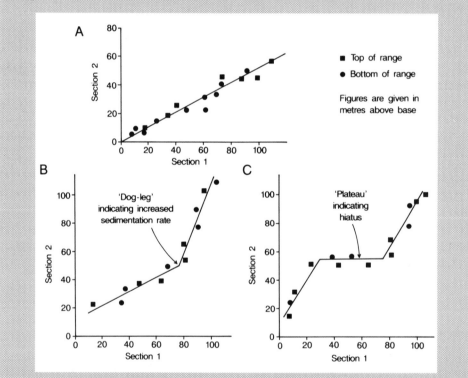

Source: Shaw, A. B. 1964. *Time in Stratigraphy*. McGraw-Hill, New York. [Diagram
   modified from: Shaw (1964) *Time in Stratigraphy*, McGraw-Hill, Figs 20.2, 20.4 &
   21.2, pp. 132, 136 & 145]

BOX 4.3: PUNCTUATED EQUILIBRIA AND PHYLETIC GRADUALISM

In 1972, two American scientists, Niles Eldredge and Stephen Gould, wrote a paper which attempted to view evolution from the perspective of the stratigraphical record. Up until then, the accepted model of evolution was that it progressed in a gradual manner with numerous intermediates, denoting evolutionary change of the entire population of a species. Eldredge and Gould called this **phyletic gradualism**. Darwin recognised that theoretically the morphological intermediates of an evolutionary line (a **phylogeny**) could be preserved in the fossil record. However, he also recognised the truth of the matter, that is, more often than not, intermediates between species (**missing links**) could not be found through the practical study of sedimentary successions. Darwin and many other scientists accounted for this by invoking the imperfection of the stratigraphical record, considered to be full of gaps. Eldredge and Gould questioned this on the basis of careful stratigraphical study. From their studies of Devonian trilobites and Pleistocene land snails, these authors asserted that in fact the stratigraphical record was a good one and that it supported rapid evolutionary change punctuated by long time intervals when there were no major changes in morphology. In doing this, Eldredge and Gould attacked the traditional idea of evolution by phyletic gradualism, and suggested that the stratigraphical record of unchanging morphology (**stasis**) followed by rapid morphological change was more illustrative of the true model of evolution, and they called this **punctuated equilibria**. This model was accepted by many geologists as representative of the evolutionary changes encountered in fossil organisms, and the debate continues unabated. Recent evidence suggests that, in some cases, evolution may actually represent a composite of both processes: **punctuated gradualism**.

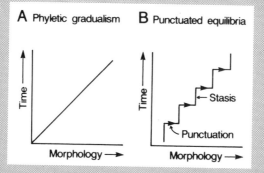

Source: Eldredge, N. & Gould, S. J. 1972. Punctuated equilbria: an alternative to phyletic gradualism. In: Schopf, T. J. M. (Ed.) *Models in Paleobiology*. Freeman, Cooper & Company, San Francisco, 82–115.

will not be detected as being time transgressive and will normally be treated as if it was formed vertically by uniform sedimentation over its area. In such cases diachronism will be undetected by biostratigraphy. However, in large depositional environments which form over much greater periods of time, biostratigraphy provides the key to the recognition of diachronism (Figure 3.12).

One of the best examples of a large-scale diachronous deposit is that of the Lower to Middle Jurassic yellow sandstones of southern England, which crop out continuously from Dorset to the Cotswolds. These sandstones have a unique colour and mineral content, forming an extremely distinctive and internally homogeneous deposit which belongs to one lithostratigraphical unit (a formation). However, detailed study of the ammonite faunas of these sandstones at a variety of localities shows the sand body to be diachronous (Figure 4.4). In the Cotswolds the sandstones contain ammonites indicative of a much older biozone than those present in the same sandstone units in Dorset. Sedimentological models suggest that these sandstones formed on a shallow shelf as a sand-bar or sheet which migrated progressively southwards in response to tidal or current activity. Although the detail of this model has been questioned, it illustrates the time-transgressive nature of a deposit which has a homogeneous lithological character (Figure 4.4).

In summary, biostratigraphy provides the key to the relative age of lithostratigraphical units. Biostratigraphical units have boundaries that are time significant and **isochronous** (of equal age), while lithostratigraphical units have boundaries based only on lithology and may therefore be diachronous.

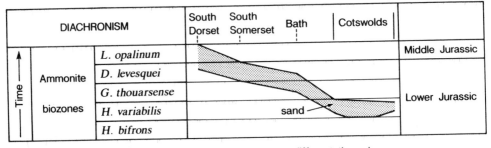

The sand horizon was laid down at different times in different regions and is said to be DIACHRONOUS

Figure 4.4 *Diachronism: an example from the Lower Jurassic sandstones of southern England. The sandstone gets progressively younger moving from the Cotswold Hills to Dorset, as shown by the ammonite biozones. [Modified from: Rayner (1981)* The Stratigraphy of the British Isles, *Cambridge University Press, Fig. 61, p. 81]*

4.2 EVENTS IN THE GEOLOGICAL RECORD

The sedimentary record is often thought to have been formed by gradual accumulation over a significant period of time. Closer examination of present-day processes and environments shows, however, that some deposits are actually deposited relatively quickly and may be considered as instantaneous when measured against the vast scale of geological time. These deposits can often be related to specific depositional **events**, such as a volcanic eruption or a severe storm. By their nature these **event horizons** are time significant and important as an independent means of providing relative dates for the stratigraphical succession.

James Hutton's principle of the uniformity of process, now termed actualism, allows geologists to deduce the nature of the agencies involved in the production of a lithology from direct observation of similar deposits which are forming today. In the 1930s it was discovered that submarine canyons were carved in the slopes of continental shelves as a result of catastrophic flows of sediment. For example, on 18 November 1929, at Grand Banks, Newfoundland, one such flow sequentially cut transatlantic telephone cables over a period of 13 hours, which allowed the velocity of the flow to be calculated at over 45 knots. This is a dramatic event, which geologically speaking was instantaneous and which deposited a large body of sediment known as a **turbidite**. This deposit can be said to have been both rapidly deposited, and associated with a specific event (the turbidity current). As another example, it has been shown recently that severe storms or giant waves (tsunamis) have had a significant impact on the sedimentological record and produce deposits which are effectively created instantaneously (Box 2.1). Many other examples of catastrophic events have been recorded in the geological record and provide important time horizons.

4.2.1 Event Stratigraphy

The products of catastrophic events are geologically instantaneous and therefore effectively represent punctuations of the stratigraphical column which can be used to denote time planes. Such **isochronous horizons** are known as **event horizons** and their study is that of **event stratigraphy**.

The value of event horizons lies in the fact that they are deemed to have been produced at the same instant over the area that they were deposited. Some deposits may be genuinely instantaneous, with time scales of hours or minutes, or alternatively they may have accumulated over some time following a short-lived event, such as in the fall-out of volcanic dust after an eruption (Figure 4.5). In an eruption, the coarser particles are deposited first, while the finer fractions take time to be deposited out of suspension. However, in geological terms such time-averaging of events is negligible and difficult to detect. Geologically, therefore, such horizons (e.g. ash horizons) can be considered to be **isochronous** (i.e. the same age everywhere) and, if present over a wide area, can be used to correlate one geological section with another. In some cases it has been shown that correlation can be achieved to a greater resolution than that of the best biozones. Some event horizons, such as those produced by the fall-out of volcanic ash, have the advantage of being easily recognisable in the rock record and may cover a large area. Ash may be deposited in a wide variety of different environments (e.g. at sea, on land or in lakes) providing the potential to correlate between very different depositional environments, something which is difficult to do with fossils, because there are few fossils common to all environments (e.g. marine animals cannot live on land). **Tephrostratigraphy** is the study of

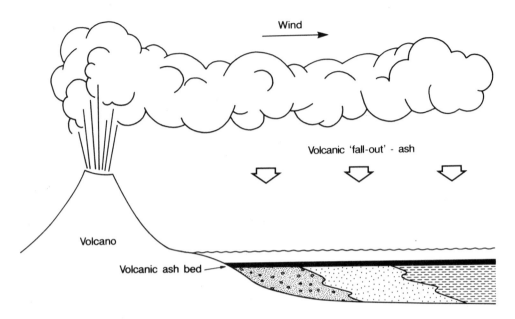

Figure 4.5 *Tephrostratigraphy: an example of event stratigraphy. The layer of volcanic ash is deposited across several different depositional environments or facies and provides an isochronous horizon*

such volcanic deposits and is widely used as a correlative tool, since each ash horizon usually has a distinct geochemical signature (Box 4.4).

Events can be classified by their origin into (1) **physical events**, (2) **chemical events**, (3) **biological events** and (4) **composite events**.

Obvious examples of **physical event horizons** are **tephra** or **ash layers** produced by volcanic eruptions. The eruption of Mount St Helens in 1981 produced large volumes of tephra which will eventually be incorporated into the stratigraphical record as a physical event horizon. The products of storms, tsunamis, meteorite impacts and mass movements all provide examples of physical event horizons. Less dramatic but more regular physical events are produced by the reversal of the Earth's magnetic field probably caused by switching of the electrical and fluid current flows in the Earth's core. These reversals impose a **magnetic fingerprint** on sedimentary and igneous rocks, because any magnetic particles within them align themselves with the prevailing magnetic field and become locked into the rock as it is lithified or as it cools (Figure 4.6). These magnetic events are a powerful tool in correlation in their own right and their study is often referred to as **magnetostratigraphy**. Many geologists consider magnetostratigraphy to be a branch of stratigraphy in its own right, but it is effectively related to specific events in Earth history and is therefore a part of event stratigraphy.

Chemical event horizons are often difficult to identify without detailed technical equipment, but usually take the form of 'spikes' in the geochemical signature of lithological units which illustrate unusual concentrations of particular stable isotopes

BOX 4.4: TEPHRA LAYERS AS CHRONOSTRATIGRAPHICAL MARKERS

Tephra layers are defined as discrete beds of volcanic ash. These are commonly found lithified as tuffs, or chemically altered into clay layers known as bentonite. Tephra layers can be extremely widespread in distribution especially if the ash has been erupted high into the Earth's atmosphere in what is known as a Plinian eruption. A Plinian eruption is exceptionally explosive and therefore injects large amounts of ash into the upper layers of the atmosphere ensuring its widespread distribution. Correlation is only possible if the layers are sufficently distinct to allow their recognition in different sections. Recognition of layers relies on petrological identification, the chemical composition (chemical fingerprint) of the individual ash or the geophysical properties. Tephra can also be radiometrically dated to give an age in years for the eruption of the material. This therefore provides the potential for accurate relative and absolute dating of sedimentary sequences.

Source: Knox, R. W. O' B. 1993. Tephra layers as precise chronostratigraphical markers. In: Hailwood, E. A. & Kidd, R. B. (Eds) *High Resolution Stratigraphy*, Geological Society Special Publication No. 70, 169–186.

(Figure 4.7). **Chemostratigraphy** is a term often used to refer to the use of such event horizons in stratigraphy. Within the Chalk rocks of southern England and central North America chemical precipitates, chemical event horizons, provide valuable information in determining the chronological sequence of these units.

Biological events are mostly associated with the rapid colonisation of particular environments, usually on a small geographical scale, or are otherwise associated with extinction events. Often, events may be **composite** in origin, combining physical, chemical and biological characteristics. For example, the eruption of Mount St Helens is obviously a physical event: the tephra has a physical presence as well as a geochemical signature and the destruction of the surrounding forest was a biological event.

Perhaps the most topical events in the geological record are **mass extinctions**. These events are effectively dramatic reductions in the diversity of animal and plant life. Mass extinctions are examples of biological events which may be caused by physical events, such as meteorite impacts. They occur over an extremely short space of geological time and many may be attributed to the catastrophic effects of changes in climate and sea level which destabilised the Earth's ecological balance. The most famous example of a mass extinction occurred at the end of the Cretaceous Period. In fact this event horizon marks the boundary between the Mesozoic and Cenozoic erathems. Recent work has attempted to explain this event in terms of a global catastrophe caused by the impact of an extra-terrestrial object, probably an asteroid, of over 10 km in diameter. The evidence for this lies in the recognition of a thin layer of clay at many sites across the globe, in which there is an abnormal enrichment in the element iridium (Figure 4.7). On Earth, iridium is concentrated in the core and is uncommon in the crust.

A

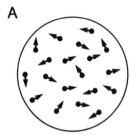

Magnetic sensitive elements randomly
orientated in magma

B

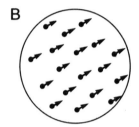

As the magma cools, magnetic sensitive elements
become aligned to the Earth's magnetic field

C

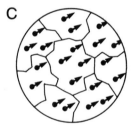

When the magma finally cools and becomes
crystalline, the Earth's magnetic field at the
time of cooling is 'locked into' the rock

Figure 4.6 *The mechanism of magnetic fingerprinting or palaeomagnetism*

Iridium is, however, common in meteorites and other extra-terrestrial bodies. This layer of clay acts both as a chemical and physical event horizon. Other explanations for this mass extinction have been proposed and are discussed in Section 10.4.1.

In summary, catastrophic events within the geological record provide important horizons or layers with which to correlate rock sequences between different areas which experience the same event. The power of the tool lies in the fact that the geological record of most catastrophic events is effectively isochronous.

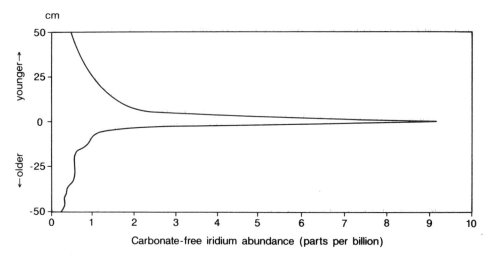

Figure 4.7 *A chemical event horizon. The element iridium is concentrated into a distinct layer, and is detected through geochemical analysis as an anomaly or 'spike'. This layer of iridium has been considered to be a product of a major meteorite impact and forms an isochronous surface or event horizon. [Modified from: Alvarez & Asaro (1992) In: Bourriau (Ed.)* Understanding Catastrophe, *Cambridge University Press, Fig. 3. p. 35]*

4.3 THE CHRONOSTRATIGRAPHICAL SCALE

One of the most important goals in stratigraphy is establishing the global standard or chronology of geological units. It should be possible to correlate the rocks of given areas with this global scale, so that geologists, wherever they may be working, can locate the rocks upon which they are working within Earth history. This global scale is known as the Chronostratigraphical Scale and has been established through the work of numerous geologists. The Chronostratigraphical Scale is a summation of all stratigraphical knowledge and as such there is no single section on the surface of the Earth at which all its units are exposed. The Chronostratigraphical Scale consists of chronostratigraphical units.

Chronostratigraphical units are bodies of strata that were formed during specific periods of geological time. The boundaries of chronostratigraphical units are time significant as they are of the same age across the globe (i.e. they are isochronous). The Chronostratigraphical Scale provides the global standard with which local sequences may be correlated. A simplified version of the currently accepted scale is given in Figure 4.8. This standard allows a uniformity of approach to be achieved by geologists worldwide and is agreed through international co-operation.

Chronostratigraphical units are sometimes referred to as **time-stratigraphical units** to distinguish them from **rock-stratigraphical units** (lithostratigraphical units). Chronostratigraphical units are in essence the repository for all stratigraphical knowledge. The mechanisms of the chronostratigraphical procedure are discussed in Box 4.5. The Chronostratigraphical Scale is a standard with which geologists can correlate their

GEOLOGICAL TIME SCALE

Cenozoic

Age (Ma)	Period	Epoch
	Quaternary	Holocene
		Pleistocene
		Pliocene
5		
10	Neogene	Miocene
15		
20		
25		
30	Tertiary	Oligocene
35		
40		
45	Palaeogene	Eocene
50		
55		
60		Palaeocene
65		

Mesozoic

Age (Ma)	Period	Epoch
80	Cretaceous	Late
100		
120		Early
140		
160	Jurassic	Late
180		Middle
200		Early
220	Triassic	Late
240		Middle
		Early

Palaeozoic

Age (Ma)	Period	Epoch
	Permian	Late
		Early
300	Carboniferous	Late
		Early
400	Devonian	Late
		Middle
		Early
	Silurian	Late
		Early
	Ordovician	Late
		Middle
500		Early
	Cambrian	Late
		Middle
		Early

Precambrian

Age (Ma)	Eon	Era
	Proterozoic	Late
1000		
		Middle
1500		
2000		Early
2500		
	Archaean	Late
3000		Middle
3500		Early

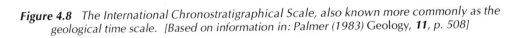

Figure 4.8 *The International Chronostratigraphical Scale, also known more commonly as the geological time scale.* [Based on information in: Palmer (1983) Geology, **11**, p. 508]

BOX 4.5: MECHANISMS OF CHRONOSTRATIGRAPHY

The basis of the concept of chronostratigraphy is that it acts as a repository for stratigraphical information. Chronostratigraphical units are bodies of rock whose boundaries are time significant, although not necessarily lithologically distinct from the units above and below. As a consequence, international agreement is needed to determine where the boundaries of global reference sections should lie and at that point a symbolic 'golden spike' is driven in. Correlation of other rock sequences with the global standard is undertaken on the basis of independent relative dating methods such as biostratigraphy or event stratigraphy, but not usually lithostratigraphy. Absolute dates may be obtained for the duration of chronostratigraphical units through the application of radiometric dating.

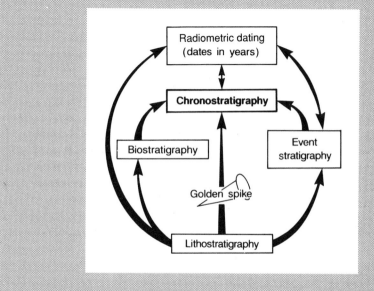

Source: Holland, C. H. 1986. Does the golden spike still glitter? *Journal of the Geological Society of London* **143**, 3–21.

individual rock sequences. For the most part the scale developed over the last 150 years and consists of a series of systems (e.g. the Carboniferous System or Triassic System). Each system consists of rocks deposited during the same time interval. Initially the systems were erected on the basis of gross lithological similarity, but for the most part they are determined on the basis of their fossil content. This allows an independent and time-significant yardstick with which to measure the extent of each system. The development of the Chronostratigraphical Scale is presented in Box 4.6. Most of the boundaries of chronostratigraphical units are determined using biostratigraphy, although other techniques are increasingly being used. Since there is no locality at which all the chronostratigraphical units which make up the global scale are exposed,

BOX 4.6: THE DEVELOPMENT OF THE CHRONOSTRATIGRAPHICAL SCALE

The development of the Chronostratigraphical Scale is seen by some scientists to be one of the most significant achievements in geology. Each of the systems was developed for the most part over a period of 50 years in the first part of the 19th century. As illustrated in the diagram, most of the systems were established from the study of the stratigraphical record in Europe and were initially defined on lithology alone. An example of a system divided on the basis of lithology is provided by the Triassic which is divided into three lithological divisions. The development of the Chronostratigraphical Scale as we know it today can perhaps be attributed to Murchison, who in 1835 defined his Silurian System on the basis of its fossil content and thereby paved the way for the recognition of major rock bodies with time-significant boundaries.

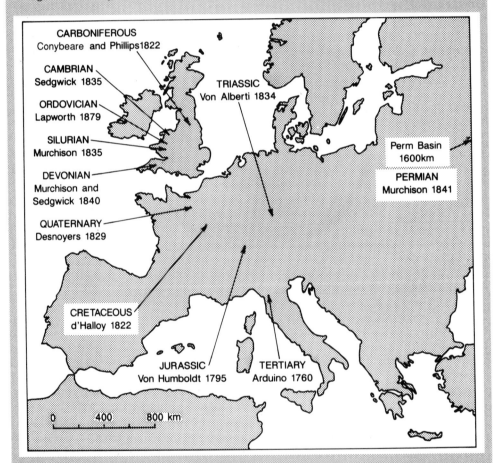

Source: Berry, W. B. N. 1986. *Growth of the Prehistoric Time Scale Based on Organic Evolution*. Blackwell, Oxford. [Diagram modified from: Eicher (1976) *Geologic Time*, Prentice-Hall, Fig. 3.2, p. 59]

individual boundaries are identified through **stratotype sections** located across the world. In other words the best exposed example of the boundary between each pair of units is selected as typical, and designated as a stratotype section, usually through international agreement. This allows geologists to correlate their individual sections with a known, protected locality which they can visit. The systems were originally defined largely in geological and geographical terms and as a consequence the boundaries were not always fixed, often causing confusion. More recently the boundaries have been decided by international agreement and a symbolic spike, known as a **golden spike**, is 'driven in' to the chosen stratotype section, to mark the agreed position of the isochronous boundary.

The systems may be grouped into larger units, known as **erathems**. These too are defined by faunal changes. John Phillips (1800–1874), the nephew of William Smith, subdivided the part of the geological record which contained fossils into three main subdivisions based on major changes in the fauna (Figures 4.9 and 4.10). These erathems

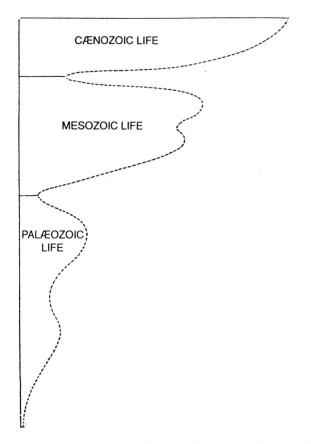

Figure 4.9 *Phillips' erathems, based on faunal diversity. The area to the left of the dashed line represents the diversity of life; the boundaries of the erathems are drawn at intervals of major falls in diversity of life. [From: Phillips (1860)* Life on the Earth: Its Origin and Succession, *Fig. 4, p. 56. Reproduced by permission of the Natural History Museum, London]*

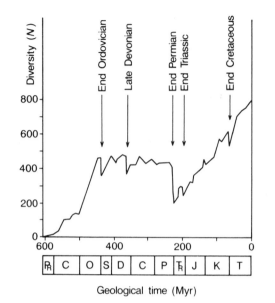

Figure 4.10 *The five major periods of extinction during the Phanerozoic. During these intervals, diversity has declined rapidly. The most dramatic event was in the Permian. The diagram also illustrates that there were many minor extinction events. [Modified from: Erwin et al. (1987) Evolution **41**, Fig. 1, p. 1178]*

have their boundaries marked by mass extinction events. Phillips named them with respect to the nature of their fauna: **Palaeozoic** (old life), **Mesozoic** (middle life) and **Caenozoic** (new life: now commonly called the **Cenozoic**), reflecting increasingly familiar fossil organisms through time to the present. The Palaeozoic–Mesozoic boundary is denoted by the Permian–Triassic mass extinction and the Mesozoic–Cenozoic by the Cretaceous–Tertiary mass extinction. The boundary between the older Primary or **Precambrian** rocks and the Palaeozoic was taken at the first appearance of shelly fossils, also now interpreted as a biological event horizon associated with the diversification of life on the planet. In this way, the major subdivisions of the Chronostratigraphical Scale are bounded by biological event horizons.

Recently it has become the practice to distinguish between chronostratigraphical units and geological time (Figure 4.11). Chronostratigraphical units are those bodies of rock deposited during a specific period of geological time. The actual time elapsed may only be measured by the application of **radiometric dating** which is discussed below. Geological time is an abstract concept and there is no guarantee that the chronostratigraphical record is truly representative of the whole of the amount of geological time elapsed (Figure 4.11). There may, for example, be periods in Earth history when no rocks were deposited anywhere. Geologists make the distinction because it is useful to have time terms (e.g. **Early** Carboniferous) to be able to refer to historical events and circumstances and chronostratigraphical terms (e.g. **Lower** Carboniferous) to refer to the relative position of stratigraphical units. The equivalent chronostratigraphical and geological time terms are given in Table 4.1.

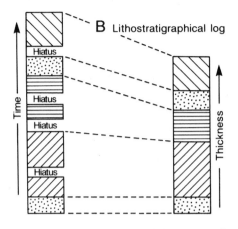

Figure 4.11 *The relationship of chronostratigraphy and lithostratigraphy. The same rock sequence is plotted in (A) with time as the vertical axis, and in (B) with thickness as the vertical axis. Thickness is not a true indicator of time elapsed, because there are periods of non-deposition (hiatuses) which are clearly illustrated on the chronostratigraphical log*

Table 4.1 *Chronostratigraphical units and their equivalent geological time units*

Chronostratigraphical terms	Geological time terms
Eonothem (e.g. Phanerozoic Eonothem)	Eon (e.g. Phanerozoic Eon)
Erathem (e.g. Palaeozoic Erathem)	Era (e.g. Palaeozoic Era)
System (e.g. Carboniferous System)	Period (e.g. Carboniferous Period)
Series (e.g. Lower Carboniferous Series)	Epoch (e.g. Early Carboniferous Epoch)
Stage (e.g. Visean Stage)	Age (e.g. Visean Age)

4.4 ABSOLUTE GEOLOGICAL TIME

Absolute dating is distinguished from **relative dating** as it deals with absolute numbers rather than the simple ordering of sequences based on fossils and other data. It relies for the most part upon the application of radiometric techniques.

The possibility of using naturally occurring radioactive elements to date rocks was first suggested by the physicist Lord Rutherford in the early 20th century. Arthur Holmes was the first geologist to construct a geological time scale based on radiometric dating and this has formed the basis of the refined high-resolution scale available today

(Figure 4.8). The method applied then, as now, is based upon the radioactive decay properties of unstable isotopes. These unstable isotopes break down to more stable ones by emitting atomic particles and/or energy. This **radioactive decay** is time dependent and forms the basis of radiometric dating.

4.4.1 The Principles of Radiometric Dating

The nucleus of an atom contains positively charged mass particles called **protons** and mass particles with no electrical charge known as **neutrons**. The nucleus of an atom occupies only about 10^{-4} of the volume of the atom but contains essentially all the mass. Orbiting the nucleus of an atom at various distances are particles known as **electrons**, the outermost of which control the chemical properties. Chemical elements are classified according to the composition of the nucleus. Two terms are used: (1) the **atomic number**, which is the number of protons within a nucleus and (2) the **atomic mass number** which is the number of protons plus neutrons. It is the convention to give the numerical value of the atomic number as a superscript and the atomic mass number as a subscript on the left-hand side of the symbol for a chemical element. For example, the element ^{238}uranium has a nucleus with 92 protons and 146 neutrons and therefore has total of 238 particles in its nucleus: $^{238}_{92}U$.

For an individual element the atomic mass number can vary, since the number of neutrons in the nucleus is not always constant. Elements having the same number of protons, but a different number of neutrons, are called isotopes. For example, oxygen has two isotopes. Both have the same number of protons (8), but one has 8 neutrons and the other 10, which gives ^{16}O and ^{18}O. Each isotope of an element is called a nuclide. Certain, although not all, isotopes are unstable and will alter through time by the emission or sometimes capture of atomic particles together with electromagnetic radiation to obtain a more stable form. This alteration is known as **radioactive decay** and the isotopes involved are known as radioactive nuclides. In radioactive decay, the original isotope is termed the **parent nuclide** and the product of the decay is called the **daughter nuclide**. The value of this process to geology is that the rate of radioactive decay is characteristically exponential and time dependent. The time scale of decay is often expressed by the **half-life** of the nuclide. This is defined as the period of time required to reduce a given number of parent nuclides by one half, and this is constant for each isotope. For example, if 100 atoms of a parent nuclide are left to decay, after one half-life only 50 of the parent atoms will remain. It will then take the same period of time to reduce the remaining 50 parent atoms to 25, and so on.

Given these principles, an absolute radiometric date can be obtained from a specific radioactive nuclide so long as the following information is obtained: (1) the proportion of the parent nuclide present; (2) the proportion of the daughter nuclide present; and (3) the rate at which decay occurs or the half-life. Given these three pieces of information an absolute age in years for a rock can be obtained. Ideally for highly accurate age dating certain conditions should be met: (1) when a radioactive nuclide first forms or becomes incorporated into a rock none of its non-radiogenic nuclides must be present and (2) no parent or daughter nuclides must be added or removed from the rock being dated, in other words the system must be a closed one. In practice these ideal conditions rarely occur but deviation from them can be readily corrected for, enabling the calculation of quite accurate absolute age estimates.

4.4.2 Methods and Limitations of Radiometric Dating

A variety of different radioactive nuclides can be used for dating geological events and Table 4.2 gives some of the most common decay series used and their respective half-lives. The magnitude of the half-life gives some idea of the length of time over which each method is appropriate: the shorter the half-life the shorter the time period over which reliable data can be obtained. For example, ^{14}carbon (commonly called carbon 14) dating can only be used to provide dates during the later part of the Quaternary as its half-life is only 5730 years; in contrast ^{238}uranium has a half-life of 4469 million years and can therefore be used to date Precambrian rocks.

Table 4.2 *A selection of radioactive reactions used in dating the geological record*

Parent: starting point	Daughter: product	Half-life (million years)
^{14}Carbon	^{14}Nitrogen	0.005 73
^{87}Rubidium	^{87}Strontium	48 000
^{40}Potassium	^{40}Argon	11 930
^{232}Thorium	^{208}Lead	14 000
^{235}Uranium	^{207}Lead	704
^{238}Uranium	^{206}Lead	4 469
^{147}Samarium	^{143}Neodymium	106 000

One limitation of the method is that it is almost totally restricted to dating crystalline rocks (igneous or metamorphic). Both these rock types are relatively closed systems, in which the radioactive clock is set at the time of crystallisation of their constituent minerals. In contrast, with very few exceptions, any radiometric date derived from sedimentary rocks will give only the ages of the source rocks from which their minerals were derived unless the mineral itself formed at the time of formation of the sediment. Carbon 14 dating of organic matter in sediments is the major example of these exceptions. The unstable carbon 14 nuclide is constantly being formed in the upper atmosphere through the bombardment of the stable nitrogen 14 nucleus by cosmic rays. This radioactive carbon rapidly cycles through the atmosphere into the hydrosphere and the biosphere. Once locked into sediments in the form of organic material it behaves essentially as a closed system, its clock starting from the time of incorporation in the sediment. With a half-life of only 5730 years, the nuclide can only be used to date quite young rocks (i.e. Late Quaternary). Sedimentary sequences may be dated, however, through the use of age **bracketing**. For example, in a sequence of sedimentary rocks which are interbedded with lava flows it is possible to date the lava flows and use them to provide age brackets for the sedimentary rocks between them. The technique of age bracketing is common when dating Quaternary sediments and is illustrated in Box 4.7.

Radiometric age dating of the constituent minerals of metamorphic rocks can provide accurate estimates of the timing of the metamorphic event, since the minerals were formed at that time. Metamorphosed igneous rocks can however provide additional important information on their original igneous age. Dating of the minerals will clearly give the age of metamorphism, but dating of the whole rock should give the original igneous age, assuming the system has remained closed. These differing techniques are known as **mineral age dating** and **whole rock age dating**.

4.5 SUMMARY OF KEY POINTS

- Biostratigraphy is the application of fossils to stratigraphy, in particular as an independent means of correlation of lithostratigraphical sequences.
- Fossils useful in biostratigraphy are known as **guide** or **index fossils**. Not all fossils are appropriate as guide fossils. Guide fossils should be **independent of environment, fast evolving, geographically widespread, abundant, readily preserved** and **easily recognisable**.
- The fundamental unit of **biostratigraphy** is the **biozone**, a body of strata distinguished by its fossil content.
- **Biostratigraphical units** are fundamentally **time significant**, independent of lithology. **Lithostratigraphical units** are not inherently **time significant** and are dependent on lithology.
- **Event stratigraphy** is the recognition of the products of geologically instantaneous events in the geological record and the use of them in correlating sequences. **Event horizons** are **time significant** and **isochronous**.
- **Events** may be **physical** (e.g. volcanic eruptions), **chemical** (e.g. regional change in water chemistry), **biological** (e.g. extinctions) or **composite**.
- **Chronostratigraphical units** are bodies of strata that are bounded by **isochronous surfaces** and which are therefore **time significant**. **Chronostratigraphical units** include the **systems** (e.g. Carboniferous System), defined mostly on biostratigraphical grounds. **Event horizons** provide an additional means of defining chronostratigraphical unit boundaries.
- The **Chronostratigraphical Scale** of chronostratigraphical units exists as a means of international correlation.
- The dating of geological events in years is known as **absolute dating**. Absolute dating relies on the known radioactive decay of certain unstable isotopes.

4.6 SUGGESTED READING

Eicher's book (1976), although quite old, provides a very nicely written and readable account of the principles of time and correlation in stratigraphy: Chapter 6 contains a

BOX 4.7: RADIOMETRIC AGE BRACKETING: AN EXAMPLE FROM THE BRITISH QUATERNARY

Radiometric dates within the recent Quaternary record are mainly based on carbon 14 dating. Evidence of a glacial advance is recorded by the presence of a glacial till; unfortunately a glacial till cannot be dated directly. It can, however, be dated through age bracketing. As a glacier or ice sheet advances to deposit the till, organic remains may be overidden and collect in pockets beneath the till layer. Similarly, as the glacier decays, organic material may collect in hollows on the surface of the till. Consequently, at exceptional locations one may find a till which is bracketed top and bottom by organic remains which can be dated by radiocarbon assay. The dates from the organic material above and below the till provide an upper and lower age for the till layer and therefore provide an estimate of the duration of the glacial advance. The best age control is provided by dating the topmost layer of the basal organic material and the bottommost layer of the organic material above the till.

Age bracketing has been used at Dimlington in Holderness, north-east England, to provide age limits for the main glacial advance of the Devensian glaciation in that area. The Devensian glaciation was the last main ice advance of the Quaternary 'Ice Age' to cover much of the British Isles. The age constraint provided at Dimlington is sufficiently good for this site to be regarded by most scientists as the type section for this Quaternary glacial advance.

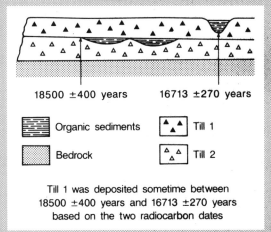

18500 ±400 years   16713 ±270 years

▦ Organic sediments   ▲ Till 1

☐ Bedrock   △ Till 2

Till 1 was deposited sometime between
18500 ±400 years and 16713 ±270 years
based on the two radiocarbon dates

Sources: Lowe, J. J. & Walker, M. J. C. 1984. *Reconstructing Quaternary Environments.* Longman, London. Rose, J. 1985. The Dimlington Stadial/Dimlington Chronozone: a proposal for naming the main glacial episode of the Late Devensian in Britain. *Boreas* **14**, 225–230.

clear discussion of the basis of radiometric dating. Snelling (1987) provides a useful discussion of the chronological subdivision of the geological record. Whittaker *et al.* (1991) provides clear definitions of the biozones and of biostratigraphy and their relationship to event stratigraphy and chronostratigraphy. Kauffman and Hazel (1977) is a collection of advanced papers on biostratigraphy while Kauffman (1988) provides an in-depth review of the concepts behind event stratigraphy. Hailwood and Kidd (1993) is a collection of rather advanced research papers on the use of various methods of determining relative time in stratigraphical sequences.

Eicher, D. L. 1976. *Geologic Time.* (Second Edition) Prentice-Hall, Englewood Cliffs, New Jersey.
Hailwood, E. A. & Kidd, R. B. (Eds) 1993. *High Resolution Stratigraphy.* Geological Society Special Publication No. 70, Geological Society, London.
Kauffman, E. G. 1988. Concepts and methods of high-resolution event stratigraphy. *Annual Reviews of Earth and Planetary Science* **16**, 605–654.
Kauffman, E. G. & Hazel, J. E. (Eds) 1977. *Concepts and Methods of Biostratigraphy.* Dowden, Hutchinson & Ross, Stroudsburg.
Snelling, N. J. (Ed.) 1987. *The Chronology of the Geological Record.* Geological Society Memoir No. 10, Geological Society, London.
Whittaker, A., Cope, J. C. W., Cowie, J. W., Gibbons, W., Hailwood, M. R., House, M. R., Jenkins, D. G., Rawson, P. F., Rushton, A. W. A., Smith, D. G., Thomas, A. T. & Wimbledon, W. A. 1991. A guide to stratigraphical procedure. *Journal of the Geological Society of London* **148**, 813–824.

5
Interpreting the Stratigraphical Record

So far we have examined the tools that allow us to determine the **stratigraphical sequence** and **age** of rock units. In this chapter we will now examine the tools that allow us to interpret stratigraphical units as the product of the environment in which they formed. From this we can make deductions about the nature of the Earth's environment through geological time.

5.1 FACIES AND THEIR INTERPRETATION

Sedimentary facies are bodies of sedimentary rock that are the product of a particular depositional environment. We can often determine this environment by examining a set of characteristics, which taken together are indicative of that environment. Facies is not a concept which is exclusive to sedimentary environments, but can also be applied in metamorphic terrains. Here the sum total of the characteristics of the metamorphic rock type allows geologists to determine the conditions existing during its formation (Box 5.1).

The concept of sedimentary **facies** was first described by Armanz Gressly (1814–1865) in 1838 while working on the Jurassic rocks of Switzerland. Gressly was concerned for the most part with the relationship between changes in fossil content and variation in lithology. He was convinced that certain fossils were linked to certain rock types and that together fossils and lithology could help determine the nature of the environment in which the rocks were deposited. Many other geologists were making similar observations and inferences at this time, but it was Gressly who first coined the term facies to encompass the sum total of all the characteristics of a rock body which allow one to determine its origin.

Actualism is the theoretical basis for facies analysis and it entails the detailed study of modern environments, to determine not only the processes by which deposition

BOX 5.1: BARROW ZONES

The application of the facies concept to mapping metamorphic rocks was explored by Barrow (1893) in a pioneering stratigraphical study of the south-east Grampians of Scotland. Within this area fine-grained metasedimentary (pelitic) rocks of the Dalradian sequence crop out. Barrow mapped a series of boundaries within these rocks on the basis of the first appearance of certain characteristic metamorphic minerals. He identified five metamorphic 'zones' of pelitic rock by this technique: chlorite, biotite, garnet, kyanite and sillimanite. The extent of each of these zones is shown in the diagram (opposite). In doing this, Barrow was simply applying stratigraphical techniques to mapping metamorphic rocks which showed changing mineral assemblages. Later experimental work showed that the sequence of zones represented progressively increasing pressure and temperature conditions, controlled by depth of burial during regional metamorphism (Chinner 1966). The work of Barrow and others therefore provides an insight into the environmental conditions of metamorphism and consequently the tectonic history of the region.

Sources: Barrow, G. 1893. On an intrusion of muscovite–biotite gneiss in the south-eastern Highlands of Scotland and its accompanying metamorphism. *Quarterly Journal of the Geological Society of London* **49**, 330–335. Chinner, G. A. 1966. The distribution of pressure and temperature during Dalradian metamorphism. *Journal of the Geological Society of London* **122**, 159–186. [Diagram modified from: Winchester (1974) *Journal of the Geological Society of London,* **130**, Fig. 3, p. 514]

occurs but also the products of those processes. In the present day it is possible to observe processes which result in a product and the causal link between the two can be established. In the stratigraphical record we have only the product. Analysis of the characteristics of that product and comparison with the products of modern depositional environments is the key to their correct interpretation. This process is the basis for most of the deductive reasoning behind the study of facies (Figure 5.1).

The characteristics of individual facies can be determined under four headings: (1) **geometry**, (2) **lithology**, (3) **sedimentary structures** and (4) **fossils**.

The geometry of a sedimentary rock body is a function of the environment in which it was deposited. River channels, sand dunes and beaches all have a characteristic shape and the sedimentary rock bodies formed in these environments will therefore have a similar geometry.

Lithology is extremely important, and careful observation of all lithological characteristics is necessary in order to deduce the type of environment in which a sediment was deposited. Of particular importance are colour, texture (i.e. grain size, grain shape and grain roundness) and composition. In desert environments, for example, iron within sediments is exposed to the atmosphere and is oxidised, which creates iron oxides (i.e. it 'rusts'). Oxidised iron minerals are known as haematite, a mineral characterised by its dark red colour. In this way red sedimentary bodies, known as **red**

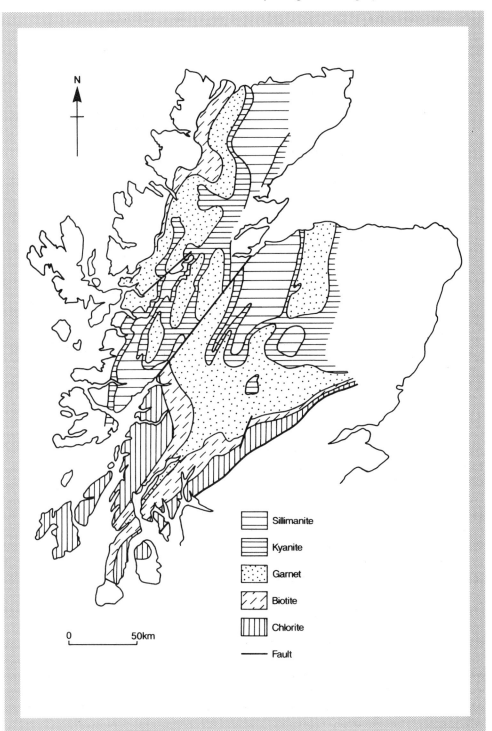

	Sillimanite
	Kyanite
	Garnet
	Biotite
	Chlorite
——	Fault

N

0   50km

PRESENT   PAST

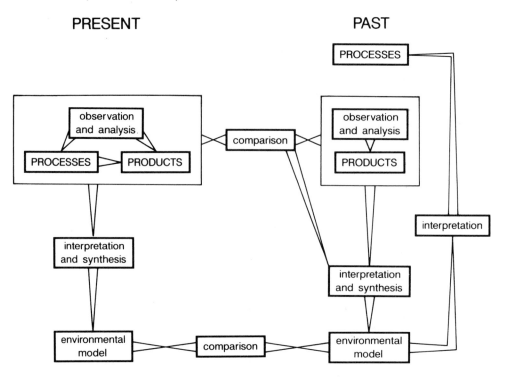

Figure 5.1 *Flow diagram illustrating the logic behind facies interpretation. [Reproduced by permission from: Earth History 1, Science Foundation Course Unit 26, Open University, p. 6]*

beds, are commonly indicative of subaerial, arid, environments. **Texture** provides information about the 'energy' of the processes which have transported and deposited the sediment. For example, fast flowing 'high-energy' rivers can transport large cobbles and boulders, while a slow moving 'low-energy' river can only transport and deposit fine-grained sands and muds. **Composition** is an indicator of the nature of the environment. For example, shelly limestones are commonly found forming today in warm, shallow marine environments, because shelly organisms thrive in such conditions. Similarly, sandstones rich in the mineral feldspar (arkosic sandstones) are indicative of an environment in which there has been little weathering, since feldspars are easily broken down into clay minerals by chemical weathering.

Sedimentary structures are extremely important in helping to deduce the processes by which different facies form. The most reliable structures are syndepositional structures, or those which formed as a direct result of the deposition of the sedimentary rock. They can provide direct evidence of whether the environment was one dominated by wind or water and can provide information on the direction of currents and the velocity of the fluid medium (wind or water) which deposited the sediment. For example, structures such as cross-bedding (Figure 5.2) indicate the nature of the environment, the direction of current flow and are also useful way-up criteria.

Fossils were important in the initial recognition of the concept of facies and are still important in the recognition of facies in the stratigraphical record. On the crudest scale, through the basic application of actualism, fossils which closely resemble the marine organisms of the present day, such as corals, can be used to indicate ancient marine environments. In much the same way, study of the interaction of present-day organisms with their environment (i.e. ecology) can be used to interpret the relationship of ancient organisms with their environment (i.e. palaeoecology).

Facies are defined on the basis of these four major characteristics and their environmental interpretation is developed through the study of modern sedimentary environments and products. Individual facies once recognised are named with regard to their **product** (e.g. sandstone facies), **process** (e.g. turbidite produced by a turbidity current) or **environment** (e.g. marine facies).

5.1.1 Facies Relationships

The relationship of one facies to another gives us the ability to assemble a picture of the geographical development of an area through geological time.

Sedimentary deposits do not continue infinitely in any one direction, but often grade laterally into other deposits or lithologies. This was recognised early in the development of geology. The use of fossils in determining relative chronology confirmed that adjacent but different sedimentary rocks were of the same age. It was apparent from Gressly's work that one adjacent facies might grade or **interfinger** (interdigitate) with another indicating the physical or geographical relationships between the original depositional environments.

One concept which is of particular importance in understanding facies relationships was introduced by Johannes Walther in 1894. Walther studied recent sedimentary facies and their relationship to the environment in which they were deposited. From these observations he deduced that environments were not static through geological time. For example, rivers change their course, even on a human time scale, and consequently deposits produced by rivers should reflect such changes. Walther noted that as environments shift position in time, the respective sedimentary facies of adjacent environments will succeed each other in a vertical profile. Therefore, in sequences where there is no apparent break in the sedimentary record the vertical profile of sedimentary facies is equivalent to the lateral variation of facies at any one time (Figure 5.3). To put this crudely, if one was to turn a vertical profile on its side then it would give a picture of the lateral variation in the depositional environments present, during the period of time represented by the vertical profile. This principle has been used to interpret vertical profiles of sediments, such as one might encounter in an exposure or borehole, as the lateral variation of sedimentary facies present within an area at that time.

Walther's principle is very valuable in situations where the relationship between facies is a purely sedimentological one, as opposed to a relationship caused by, for example, a rise or fall of sea level. Facies relationships which are purely sedimentological and are not the result of external variation (i.e. sea level) are known as **autocyclic**

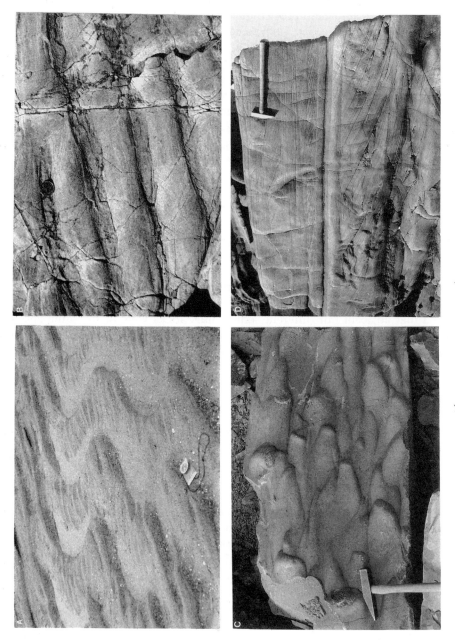

Figure 5.2 (*Caption on facing page*)

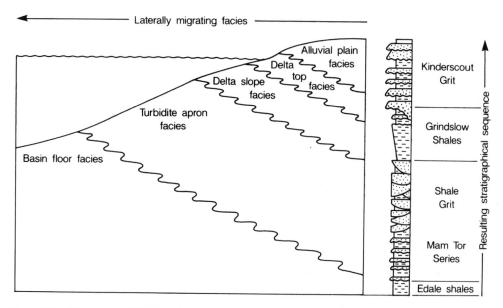

Figure 5.3 Illustration of Walther's principle. This diagram illustrates the principle that the lateral migration of sedimentary facies results in a vertical sequence of the same facies. [Modified from: Selley (1970) Ancient Sedimentary Environments, *Chapman & Hall, Fig. 5.6, p. 86*]

Figure 5.2 Sedimentary structures: selected examples. **A:** Large ripples on the surface of a beach in south Devon, formed by water flowing from left to right [Photograph: M. R. Bennett]. **B:** Large fossil ripples on a bedding plane within a series of fine-grained sandstones [Photograph: D. G. Helm]. **C:** Flute casts on the basal surface of a unit of Llandovery siltstone and sandstone from South Wales. As a turbidity current flows over a surface of sediments, turbulent flow within it produces a series of scour structures which are deeper up-flow and flare outwards in the down-flow direction. The subsequent deposition of sediment on top of the scoured surface produces a cast of the scour marks, like a plaster cast of a footprint. It is this cast which is shown in the photograph. Flute marks can be used to indicate the direction in which the turbidity current flowed, in this case from right to left [Photograph: BGS]. **D:** Cross-bedding preserved in the Moine rocks of Morar, north-west Scotland (metasedimentary rocks). These structures were formed by a small sub-aqueous dune in a shallow marine environment. Sediment was moving from left to right. On such a dune sediment first moves up the gentle stoss side of the dune; some of this sediment is deposited in thin layers known as top sets, but most of the sediment avalanches over the steep lee face of the dune (down-flow) to form steeply dipping foreset beds. These foreset beds grade into bottom set beds, which approach the horizontal surface beneath the dune asymptotically. In the photograph one can see only foresets and bottom sets , the top sets having been removed by erosion, a common occurrence, as shown by the horizontal surface which truncates the top of the foresets. Cross-bedding like this can be used, therefore, not only to indicate the direction of sediment transport, but also as a way-up structure [Photograph: BGS]

and usually reflect changes in the sedimentary environment. The gradual lateral movement of delta lobes as a response to increased sedimentary input provides an example of an autocyclic relationship.

5.1.2 Relative Sea Level, Facies and 'Sequence' Stratigraphy

In the marine environment **sea level** has a profound effect on the spatial distribution and depositional environment of sedimentary facies.

Sea level has varied throughout geological time. The basic information that allows geologists to infer this is the pattern of marine, coastal and non-marine facies. Consequently, in addition to the **autocyclic mechanisms** of the self-contained sedimentary system, controlled largely by sediment supply, the position and nature of facies may be controlled by external influences such as sea level or tectonic uplift. Such mechanisms are known as **allocyclic**. Often it is a difficult task to be able to recognise which (autocyclic versus allocyclic) has been of more importance in the development of a particular sedimentary sequence.

The relationship of facies through time can be used to deduce relative sea level variations from the stratigraphical record. The relationship of marine facies overlying non-marine or basement sequences shows that at some point sea level has risen to allow the sea to progressively **transgress** onshore. The marine facies are said to **onlap** the land surface and it is the extent of this **coastal onlap** that allows the relative rise or fall of sea level to be deduced. In some sequences the progressive landward movement of marine facies means that successively younger sets of facies are seen to **overlap** onto the land surface beyond the rock produced by a lower sea level stand. In cases where the onlap is taking place over a folded, dissected basement, the term **overstep** is used. In this case the onlapping sedimentary facies are seen to step over the exposed, tilted strata of the older sequence. This, of course, leads to the production of an angular unconformity and it is now realised that most major unconformities are produced by such successive onlap (see Section 3.3.2).

Transgressive facies patterns are produced where successive landward onlap of marine sequences takes place during a rise in sea level. In a marine setting the different environments and their facies are usually in equilibrium, so that moving offshore we might encounter the following facies succession: a beach pebbles facies—a nearshore sand facies—an offshore mud facies. In a transgressive sequence these environments and their facies will simply move progressively inshore, building a vertical sequence which obeys Walther's principle (Figure 5.4). With falling sea level, a **regressive facies pattern** is produced with the same facies pattern moving progressively offshore. The movement of facies offshore is referred to as **offlap**. An important principle to remember in transgressive and regressive facies patterns is that boundaries between facies are usually diachronous and that they cross time lines as they move progressively on or offshore (Figure 5.4). In some cases it is possible to relate transgressive and regressive facies patterns so that the record of interdigitated facies produces a zig-zag pattern such as is seen in the Cenozoic sediments of south-east England (Box 5.2), or in the Upper Mesozoic sediments of the western interior of North America.

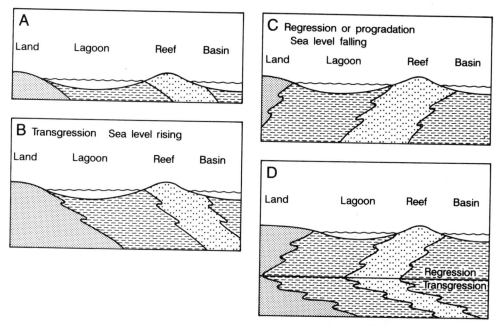

Figure 5.4 *Facies patterns caused by trangressions and regressions of the sea.* **A:** *Distribution of facies.* **B:** *The facies pattern caused by a rise in sea level.* **C:** *The facies pattern caused by a fall in sea level.* **D:** *The facies pattern caused by a trangression and regression of the sea.* [*Modified from: Stanley (1986)* Earth and Life through Time, *Freeman, Fig. 5.9, p.125*]

It is apparent from the above discussion that facies relationships provide a tool with which patterns of sea level variation through geological time can be inferred. Sea level variation may involve either (1) local or regional variations induced by local changes in land elevation (**relative sea level**) or (2) global or **eustatic** variations in sea level induced by changes in the volume of the world's ocean basins or by changes in the total volume of water present. Determining the pattern of **eustatic sea level** variation through geological time is clearly of some importance. There are two methods by which a eustatic sea level curve can be deduced. The first method is based on estimates of the **area of continental flooding**, that is, the continental area covered by marine facies during a given period of geological time. This is done by plotting the distribution of marine facies of a given age on a palaeogeographical map of similar age. This type of approach has been used by Hallam to produce a eustatic sea level curve for the Phanerozoic (Figure 5.5). The second method involves the use of **relative onlap** or **offlap**.

Stable continental margins contain a remarkably detailed sedimentary record of sea level variation. However, these sediments can only be analysed by **seismic exploration**, which employs a technique similar to echosounding. If a sound signal is sent towards the sea floor and the time taken for it to return is recorded (i.e. the echo time) then the depth of water can be determined if the velocity of sound in water is known. If a more powerful source of energy with a lower frequency is used then the energy

BOX 5.2: FACIES PATTERNS AND SEA LEVEL: AN EXAMPLE FROM SOUTH-EAST ENGLAND

The Palaeogene strata of the Hampshire and London Basins in south-east England contain a remarkable record of sea level change. In these basins there was a competitive interplay between marine and non-marine environments. Stamp (1921) set up a simple model for a sea level cycle of transgression and regression which he called the 'cycle of sedimentation'. The facies which would result from this cycle were viewed as wedge-shaped incursions of marine deposits (Diagram A, opposite). Several sea level cycles would therefore produce a facies pattern which resembles a series of zig-zags or wedges. Stamp (1921) applied this simple model to the Eocene deposits of the Hampshire and London Basins and summarised his results in a series of schematic diagrams such as the one shown in Diagram B. Stamp's work was refined by Curry (1965), although the same zig-zag pattern of facies was identified (Diagram C). The sea level changes recorded in south-east Britain are probably eustatic in nature and may be explained in terms of variations in the rate of sea floor spreading in the opening of the Atlantic.

Sources: Stamp, D. 1921. On cycles of sedimentation in the Eocene strata of the Anglo-Franco-Belgian Basin. *Geological Magazine* **108**, 108–114, 146–157, 194–200. Curry, D. 1965. The Palaeogene beds of south-east England. *Proceedings of the Geologists' Association* **76**, 151–173. [Diagram modified from: Stamp (1921) *Geological Magazine* **108** Fig. 1, p. 108; Curry (1965) *Proceedings of the Geologists' Association* **76**, Fig. 3, p. 167]

will penetrate the sea floor and be reflected from the boundaries between sediment layers. These boundaries are commonly referred to as **seismic reflectors**, or just **reflectors** for short. A seismic profile of the sediments beneath the sea floor can be obtained from seismic exploration. Such a profile resembles a geological cross-section, although the horizontal dimension records the ship's movement and the vertical scale records echo time rather than depth. The reflectors are commonly spaced somewhere in the region of 10 metres apart and are therefore unlikely to be truly representative of the full stratigraphical sequence. However, when used in conjunction with boreholes and sediment cores a detailed picture of the rocks and sediments which underlie a section of sea floor can be obtained. Most of the reflectors which return echoes, and are therefore recorded in seismic profiles, represent surfaces which were once the sea floor. As a consequence they possess a full suite of onshore–offshore facies along their length and are effectively facies-independent time planes. Each former sea floor and the sediment associated with it may either onlap or offlap previous surfaces or sea floors (Figure 5.6). By determining the pattern of relative onlap or offlap one can obtain a record of the marine transgressions and regressions which have taken place in the past.

Seismic reflection studies of continental margins have shown that the pattern of onlapping and offlapping surfaces can be grouped into sediment packages, known as '**sequences**'. Each 'sequence' is usually bounded, top and bottom, by an unconformity.

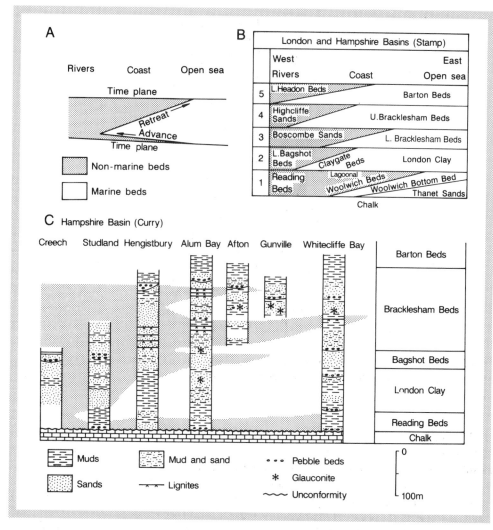

The onlapping or offlapping surfaces within a 'sequence' are known as **parasequences**. Stable continental margins often possess a stack or succession of 'sequences' produced by the transgression and regression of the sea. Since the term sequence is often used in stratigraphy in a loose sense we distinguish it, therefore, in the formal sense by placing the word in inverted commas. 'Sequences' develop at the continental margin as it is here that the major sedimentary bodies, such as deltas, form. This contrasts with the centre of the basin where sedimentation only occurs by the settling out of fine-grained sediment from suspension and there is little input of coarse sediment. Given a fall in sea level, a delta will build out seawards, or **prograde**, into the available space (the **accommodation**) of the sedimentary basin. Deltas and other major sedimentary systems are in equilibrium with sea level: that is they change their location in sympathy with changes in sea level.

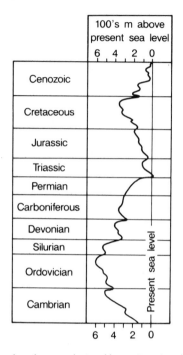

Figure 5.5 *Hallam's eustatic sea level curve derived by estimating the area of continental flooding. [Modified from: Hallam (1984) Annual Reviews of Earth and Planetary Science* **12**, *Fig. 5A, p. 220]*

The concept of sediment packages or 'sequences' along continental margins is illustrated in the block diagrams in Figure 5.7. These diagrams depict the pattern of sediment packages associated with a hypothetical continental margin subject to sea level fluctuations. In the first block diagram an initial wedge of sediment is seen to build up during a period of high sea level (Figure 5.7A). In the second block diagram a fall in sea level causes the erosion and truncation of this sediment wedge and the deposition of a sediment fan offshore (Figure 5.7B). The initial sedimentary wedge is truncated by erosion and an unconformity is formed which acts as an upper boundary to the initial, high sea level sediment wedge. Sea level has now stabilised at a low point in the third block diagram and a new sediment package builds up. Finally as sea level rises to its former level a new sediment package accumulates on top as a delta, or similar sediment body, builds up (Figure 5.7C–E). When viewed in an interpreted seismic section the sediment packages along this coastal margin should resemble those in Figure 5.7F. Two sediment 'sequences' can be recognised separated by the unconformity formed during the sea level low (Figure 5.7G).

We can explore the concept of 'sequence' development further with reference to the hypothetical continental margin seismic profile which we have interpreted in Figure 5.8. Figure 5.8A is an idealised section through a basin margin and shows the pattern of former sea floors (seismic reflectors) as well as the pattern of facies. Notice that the

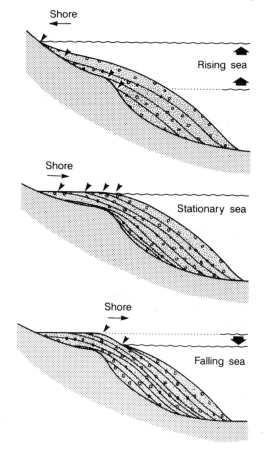

Figure 5.6 *Onlap and offlap of sedimentary bodies. The upper illustration shows how rising sea level produces a pattern of onlapping sedimentary bodies, in which each unit, indicated by the arrows, extends further onshore than the last. The middle diagram shows what happens when sea level is stationary, in which each unit is displaced seawards (offlapping) by the building out of a sedimentary body. The lower illustration shows how a fall in sea level also produces a pattern of offlapping sedimentary bodies, in which the extent of each unit is further offshore than the last. [Reproduced by permission from: van Andel (1985) New Views on an Old Planet, Cambridge University Press, Fig. 10.4, p. 152]*

facies boundaries are diachronous and along each time plane or former sea floor the same succession of facies is present. Figure 5.8B shows the main unconformities or 'sequence' boundaries present within the section. Each of these unconformities was formed by a fall in sea level. These boundaries define three 'sequences'. Figure 5.8C shows both the 'sequence' boundaries and the former sea floor surfaces which split each 'sequence' into parasequences. The relative coastal onlap and offlap of the parasequences can be seen clearly in this section. The history of this margin can be reconstructed as follows.

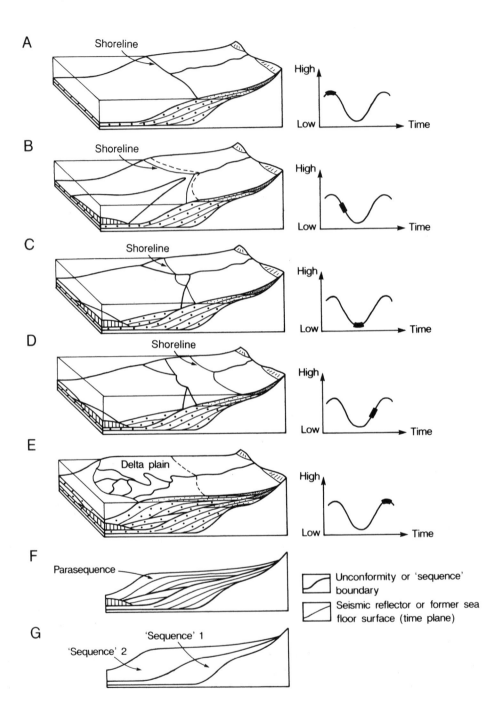

1. Parasequences 1 to 3 were deposited during a rise in sea level, the magnitude of which is illustrated by the amount of coastal onlap. Sea level reached its maximum elevation during parasequence 3. Parasequences 4 to 6 were deposited as sea level began to slowly fall, showing an offlapping relationship, which is partially obscured because of the later truncation at the 'sequence' boundary.
2. Following the deposition of parasequence 6, sea level fell rapidly. Erosion of parasequences 1 to 6 produced an unconformity and the first 'sequence' boundary. The eroded material was deposited as a small offshore fan, parasequences 7 to 8. .
3. A subsequent rise in sea level deposited parasequences 9 to 12 and again the amount of sea level rise is indicated by coastal onlap onto 'sequence' boundary 1.
4. Before sea level had risen to its former level it fell rapidly again causing erosion of parasequence 12. This produced a second unconformity and 'sequence' boundary.
5. Sea level then rose dramatically to deposit parasequence 13. Further sea level rise and progradation of the sediment bodies occurred during the deposition of parasequences 14 to 16.

This hypothetical example illustrates the principle by which 'sequences' can be recognised and interpreted in terms of sea level variation. Three 'sequences' can be recognised, each with bounding unconformities. These unconformities are produced by the action of relative sea level. By carefully measuring the amount of coastal onlap within each 'sequence' a coastal onlap curve can be produced (Figure 5.8D) from which a sea level curve can be deduced given a correction for any change in land elevation (Figure 5.8E).

It is important to note that 'sequences' are easily recognised only at basin margins. Offshore, 'sequence' boundaries (unconformities) may equate with conformities since sedimentation within the basin centre may have been continuous despite the fluctuation at the basin margin (Figure 5.9).

The study of the interrelationships of these sediment wedges or packages therefore reveals a history of local sea level change which can be obtained for any given continental margin. Correlation of local sea level histories obtained from continental margins spread across the globe allows a picture of eustatic sea level to be obtained. This type of technique has been used to obtain a eustatic sea level curve commonly referred to as the **Vail sea level curve** after the principal researcher in this field (Figure 5.10).

Consequently, it is possible to obtain a eustatic sea level curve either using the area of continental flooding (Hallam) or by using 'sequence' stratigraphy (Vail). In outline the Vail and Hallam eustatic sea level curves are very similar despite the fact that they are obtained by different methods (Figure 5.11). In detail, however, there are some significant differences between the two graphs.

Figure 5.7 *Block diagrams showing the development of sedimentary packages or 'sequences' along a continental margin with fluctuating sea level. The graphs on the right-hand side of the diagram show the position of sea level within a simple cycle of sea level change from high, through low and back to high. [Modified from: Posamentier et al. (1988) In: Wilgus et al. (Eds)* Sea Level Change: An Integrated Approach, *Society of Economic Palaeontologists and Mineralogists, Special Publication 42, Figs 1–6, pp. 111–114]*

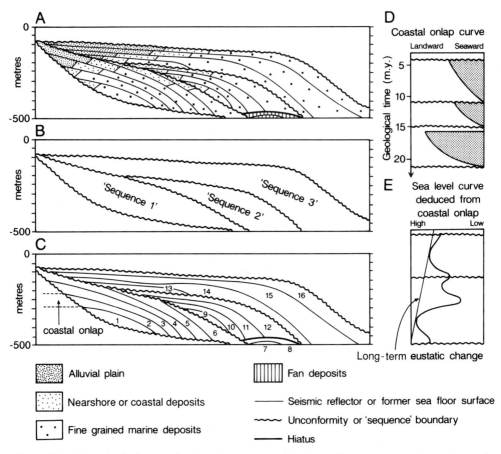

Figure 5.8 *Hypothetical example of an interpreted seismic profile across a continental margin.* ***A:*** *shows an interpreted seismic profile, showing both the facies pattern and the seismic reflectors (former sea floors) within the section.* ***B:*** *within this section the 'sequence' boundaries have been picked out, and three sedimentary 'sequences' have been identified.* ***C:*** *shows the parasequences within each of the three identified 'sequences'.* ***D:*** *shows the coastal onlap curve produced by the study of the onlap and offlap patterns within the seismic profile (C).* ***E:*** *illustrates how the coastal onlap curve can be combined with a knowledge of long-term eustatic patterns to produce a sea level curve. [Modified from: Wilson (1992). In: Brown et al.(Eds) Understanding the Earth, Cambridge University Press, Fig. 20.18, p. 411]*

The study of sea level change through sedimentary 'sequences' was first developed by oil companies as a by-product of their exploration for offshore oil reserves. More recently, however, the concept of sedimentary 'sequences' has been used as a tool in correlation both within a depositional basin and between basins (see Section 7.1.2). This application is known as **'sequence' stratigraphy** and relies simply on the recognition and correlation of depositional 'sequences' and the unconformities between them. Most 'sequence' stratigraphers argue that 'sequences' are produced in response to allocyclic mechanisms, and primarily to changes in relative sea level as we have

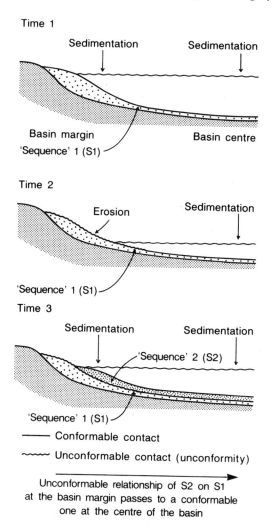

Time 1

Sedimentation

Sedimentation

Basin margin

Basin centre

'Sequence' 1 (S1)

Time 2

Erosion

Sedimentation

'Sequence' 1 (S1)

Time 3

Sedimentation

Sedimentation

'Sequence' 2 (S2)

'Sequence' 1 (S1)

——— Conformable contact

〜〜〜 Unconformable contact (unconformity)

Unconformable relationship of S2 on S1
at the basin margin passes to a conformable
one at the centre of the basin

Figure 5.9 *'Sequence' boundaries and their conformity in the basin centre. This diagram illustrates how the boundaries between 'sequences' are usually unconformities at the basin margin, becoming conformities in the basin centre.* **Time 1:** *deposition of 'sequence' 1.* **Time 2:** *a sea level fall exposes 'sequence' 1 to erosion at the basin margin, although sedimentation is continuous in the centre of the basin.* **Time 3:** *a rise in sea level results in the deposition of a second 'sequence' ('sequence' 2) which rests at the basin margin on the eroded surface of 'sequence' 1 forming an unconformity. In the centre of the basin, 'sequence' 2 rests conformably on 'sequence' 1 as sedimentation has continued uninterrupted.*

discussed above. The bounding unconformities of 'sequences' are inferred to be genetically linked to changes in sea level. If the change in sea level is eustatic (worldwide) rather than local, then it follows that the resulting sedimentary packages can be correlated on the basis of 'sequence' boundaries from one continental margin to another. Some researchers have, however, questioned whether relative uplift of continental areas or even increased

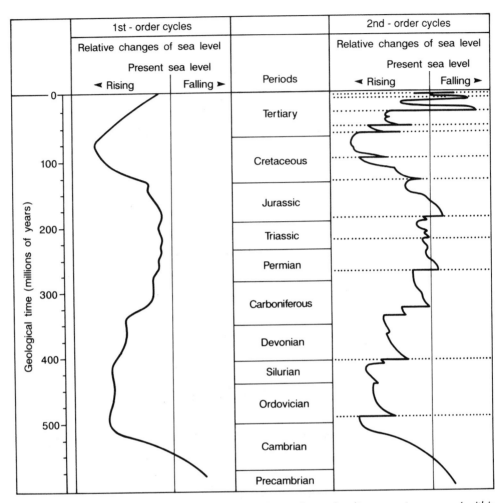

Figure 5.10 *Vail sea level curve produced from the analysis of sedimentary 'sequences' within seismic sections. [Modified from:* Vail *et al.* (1977) *American Association of Petroleum Geologists Memoir 26, Fig. 1, p. 84]*

sediment supply (an autocyclic mechanism) could not have caused the patterns of relative onlap and offlap seen within many 'sequences'. Whatever the cause of 'sequence' boundaries (allocyclic versus autocyclic), 'sequence' stratigraphers have had some success in the correlation of sedimentary 'sequences' and the unconformities which bound them.

5.2 PALAEOGEOGRAPHY

Palaeogeography is the study of the geography of the ancient Earth. On the broadest scale palaeogeography refers to the distribution of the continents and oceans. Smaller scale

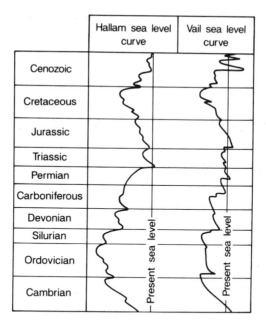

Figure 5.11   *Comparison of the Vail and Hallam sea level curves. [Modified from: Hallam (1984) Annual Reviews of Earth and Planetary Science **12**, Fig. 5, p. 220]*

geographies may be obtained through the study of facies distributions. These may ultimately be gathered together to give a detailed pattern of the geography of an area like Europe or North America through time. Local palaeogeographical maps are constructed by examining the distribution of facies of a similar age within a given area, and plotting the relative position of the different environments indicated by the facies. Such a **palaeogeographical map** might show the distribution of marine, coastal and continental environments (see Section 11.7). Palaeogeographical maps are constructed by simple facies interpretation carried out for specific periods of geological time known as **time slices** (Figure 5.12). A time slice is determined using biostratigraphy or other correlation tools which are facies independent. Once the time constraints are determined, stratigraphical sections can be correlated within a given area and the spatial pattern of facies within it at a given time can be determined. A palaeogeographical map is then constructed by interpreting each facies, so that a clear understanding of the nature and position of environments within the area is achieved (Figure 5.12). If this procedure is repeated for earlier and later time intervals then a series of palaeogeographical maps can be compared to show how an area or region has changed through time. In this way stratigraphical columns can be seen as a record of the changing geography of an area.

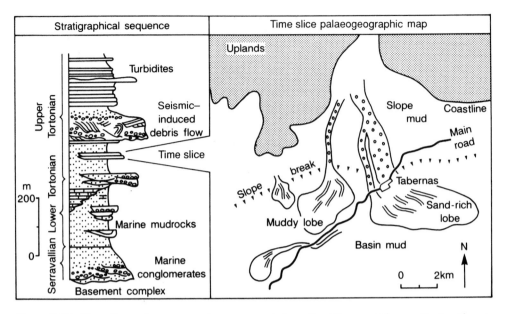

Figure 5.12 *Time slice palaeogeography, an example from the Miocene Tabernas Basin, Almeria, southern Spain. Analysis of numerous stratigraphical sections (only one is illustrated) allows sediments of the same age for a given area (a time slice) to be plotted spatially, to produce the palaeogeographical map illustrated. [Modified from: Kleverlaan (1989)* Sedimentology **36,** *Figs 2 & 6, pp. 26 & 27]*

5.3 PALAEOCLIMATOLOGY

The goal of **palaeoclimatic reconstruction** is to define the climatic conditions and characteristics of the atmosphere, oceans and landmasses through geological time. The challenge is to find geological evidence which can be used as the thermometers, barometers and anemometers of different geological periods. There are four main types of data available to reconstruct palaeoclimate: (1) **chemical data**, (2) **biological data**, (3) **physical data**, and (4) **computer simulations**.

5.3.1 Reconstructions Based on Chemical Data

The most widely used chemical method in climatic reconstruction is the measurement of **stable isotopes** (see Section 4.4). In particular, the **oxygen isotopic composition** of carbonates, secreted by organisms growing in sea water, can be used to provide palaeotemperature data. On average the ratio of ^{18}O to ^{16}O in today's oceans is about 1 to 500 or alternatively 0.2% of all oxygen is ^{18}O. This ratio is highly sensitive to temperature (Figure 5.13). Consequently, if the oxygen isotope ratio present in the calcium carbonate ($CaCO_3$) of a shell is known, an indication of the ocean temperature in which it lived can be obtained. Planktonic organisms known as foraminifera are the

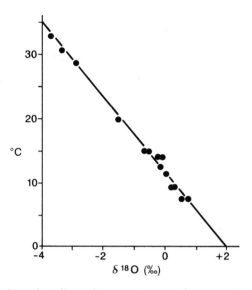

Figure 5.13 *Graph to show the effect of temperature on the proportion of ^{18}O. This graph is based on laboratory experiments conducted on the oxygen isotopic composition of molluscs growing at different temperatures and demonstrates that oxygen isotopes in carbonate shells can be used as palaeothermometers. The oxygen notation expresses the difference in the abundance of ^{16}O and ^{18}O from a standard in parts per thousand. [Reproduced by permission from: Barron (1992). In: Brown et al. (Eds) *Understanding the Earth, *Cambridge University Press, Fig. 24.1, p. 487]*

most frequently used organisms for palaeotemperature analysis because of their widespread distribution over the surface of present and former oceans.

Four factors limit the use of oxygen isotopes as a palaeoclimatic tool. First, the isotopic composition of sea water is dependent on evaporation, precipitation and run-off from land areas. Evaporation selectively removes the 'lighter' ^{16}O isotope (as 'light' water: $H_2^{16}O$) and correspondingly rain and snow is enriched in ^{16}O. If this rain or snow returns to the oceans the oxygen isotope ratio of the oceans remains unchanged, but during a glaciation, rain and snow is stored in ice sheets and does not return quickly to the oceans. Consequently, during glaciation oceans are enriched in ^{18}O (Figure 5.14). This is very useful in the study of the Quaternary 'Ice Age', since the changing ratio of ^{18}O to ^{16}O in the skeletons and shells of such organisms as foraminifera taken from ocean sediments can be used to provide a record of the growth and decay of ice sheets (Box 5.3). However, if one wants to obtain palaeotemperature estimates a correction for the presence of ice sheets is required and therefore one needs a knowledge of changes in the global ice volume. This can provide a serious limitation when considering the pre-Quaternary record, where our knowledge of the Earth's glacial history is incomplete.

The second limiting factor is that not all organisms live in balance with the isotopic composition of sea water. For example, the shells of certain organisms may have isotopic compositions which are different from that of the sea water in which they live and do not therefore record changes within it.

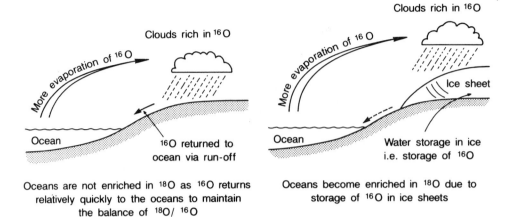

Figure 5.14 *Diagram to show the natural fractionation process which takes place during the evaporation of sea water and illustrates how the growth of an ice sheet disrupts the balance of oxygen isotopes within sea water*

Thirdly, the use of isotopic data requires a knowledge of the geographical habitat of the species being used in the analysis. For example, if the organism was mobile then this could seriously affect the interpretation of the results, the palaeotemperature estimates being an integration of the different locations or environments in which the organism lived. Finally, the isotopic composition of a shell may change during the transformation of a sediment to a rock (**diagenesis**) and may not therefore reflect the original composition of the ocean in which the shell lived. Despite these limitations the analysis of oxygen isotopes can provide important palaeoclimatic data.

The isotopes of carbon (^{12}C, ^{13}C, ^{14}C) contained within the shell of marine organisms can also be used to give an indication of the organic productivity of the oceans and the pattern of ocean circulation. This information is very important given the close coupling of the oceanic and climatic systems.

5.3.2 Reconstructions Based on Biological Data

Palaeoclimatic information can be obtained from fossils. If a certain plant or animal only lives in a specific climatic environment then its distribution can be used to provide information on the extent of that environment. This principle is well illustrated by large reptiles or amphibians which are almost exclusively tropical today. For example, crocodiles are not found poleward of latitudes with a mean annual temperature of 15°C. Consequently the presence of fossil crocodile bones within the geological record can be used to infer temperatures in excess of 15°C. In the same way hermatypic corals, which are the principal constituent of modern reefs such as the Great Barrier Reef in Australia, are largely restricted to the tropical and subtropical latitudes (less than 30° and in waters warmer than about 21°C (see Section 5.4 and Figure 5.19). Reef limestones within the geological record which contain hermatypic corals provide, therefore, a good indication of surface water temperature.

BOX 5.3: GLACIAL HISTORY AND THE OXYGEN ISOTOPE RECORD

The isotopic composition of ocean water changes with the growth and decay of ice sheets. A record of this isotopic variation and therefore a record of global ice volume is stored in the carbonate of shells within deep sea sediments. The ratio of ^{18}O to ^{16}O in today's oceans is normally about 1 to 500 and the oxygen locked up in the carbonate of marine shells reflects this ratio. This ratio varies with the growth and decay of the Earth's ice sheets. When sea water evaporates a process of natural fractionation occurs and more water molecules with ^{16}O are evaporated than those with ^{18}O because the atomic mass of ^{16}O is less (i.e. the molecule is 'lighter'). Atmospheric water, clouds and rain, are therefore enriched in ^{16}O. In a non-glacial environment the balance of ^{18}O to ^{16}O is maintained because rain water falling on land quickly returns to the ocean via rivers. In contrast, during a glacial period the oceanic balance of ^{18}O to ^{16}O is upset because atmospheric moisture is not returned quickly to the oceans but falls as snow and is stored in ice sheets (Figure 5.14). Consequently the oceans are enriched in ^{18}O during a glacial period. A record of the Earth's glacial history during the Quaternary 'Ice Age' can, therefore, be obtained from an analysis of oxygen isotopic composition of deep ocean sediments: horizons in which the shells and skeletons of marine organisms are enriched in ^{18}O correspond to periods when the Earth's total ice volume was large (**glacial periods**) while those horizons with less ^{18}O correspond to periods with a small global ice volume (**interglacial periods**).

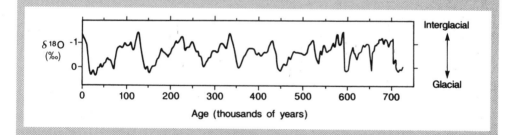

Sources: Imbrie, J. & Imbrie, K. P. 1979. *Ice Ages: Solving the Mystery.* Harvard University Press, Cambridge, Massachusetts. [Diagram modified from: Barron (1992). In: Brown *et al.* (Eds) *Understanding the Earth,* Cambridge University Press, Fig. 24.2, p. 487]

One of the major difficulties with using a given species of plant or animal as a palaeoclimatic indicator is that most organisms can survive in a range of climatic conditions. This may lead to large margins of error in any reconstruction of palaeoclimate. To avoid this problem an assemblage of faunas and floras is usually used. The overlap between the climatic preferences of one species and another helps reduce the range of possible palaeoclimates which can be inferred (Figure 5.15). This method is

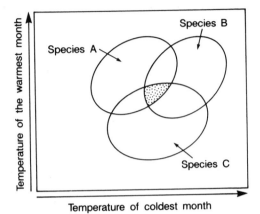

Figure 5.15 *This schematic diagram illustrates the concept of mutual climatic ranges. Three species (A, B & C) each have broad climatic preferences. A rock containing all three species as fossils is likely to have formed in a climate corresponding to the stippled area*

often referred to as the method of **mutual climatic ranges**. The application of this type of approach is well illustrated by its successful use in reconstructing the palaeotemperatures of the last 22 000 years of the Quaternary from beetle remains (Box 5.4).

In general, for a fossil to be of use as a palaeoclimatic tool it must meet the following criteria. First, it must live in equilibrium with its environment, so that any changes to that environment will lead to either adaptation, migration or mortality. Second, the principle of uniformitarianism must apply. The climatic preferences of species today must have been constant throughout their existence if they are to be used as indicators of past environments. This becomes difficult for fossil groups without living relatives. In this case the closest living relative to a fossil in either genetic or morphological terms is often used: this can, however, be misleading.

The morphology of plants can also provide palaeoclimatic information. Today different plants are adapted to different climates. For example, plants typical of a rain forest tend to possess elongated leaf end or 'drip tips' which are adapted for shedding water, while plants which live in arid areas have reduced leaf areas or have leaves which turn end-on to the sun to avoid excessive water loss through evapotranspiration. This type of adaptation is common in plants and can be recorded in plant fossils, thereby providing important palaeoclimatic data.

5.3.3 Reconstructions Based on Physical Data

Sedimentary rocks can be used to provide palaeoclimatic information, if the environment in which they were deposited was climatically controlled. The distribution through time of climate-sensitive deposits, like coals, evaporites, tillites and carbonates, have in the past been zonally distributed in much the same way as they are today (Figure 5.16). Three broad types of palaeoclimatic indicators can be recognised: (1)

BOX 5.4: THE MUTUAL CLIMATIC RANGE METHOD AND THE RECONSTRUCTION OF PALAEOTEMPERATURES DURING THE LAST 22 000 YEARS

Beetle or coleopteran remains can provide good palaeotemperature data. As a group beetles are varied and many species are climate sensitive with well-defined tolerance ranges. They are found in a wide variety of Quaternary sediments and most of the species recorded in these sediments exist today.

Atkinson *et al.* (1987) used Coleoptera to reconstruct the palaeotemperatures of Britain over the last 22 000 years. These palaeotemperature reconstructions were made by using the technique of **mutual climatic ranges**. The basic assumption is that, if the present-day climatic tolerance range of a beetle is known, then fossil occurrences of that species imply a palaeoclimate which was within the same tolerance range. If several species occur together as a fossil assemblage, then the palaeoclimate at the time they were alive must lie within the mutual intersection of their tolerance ranges. This intersection will be smaller and the palaeoclimate deduced more precise if a large number of co-existing species are used. Atkinson *et al.* (1987) first defined the climatic tolerance range of 350 living beetle species which commonly occur as Quaternary fossils. This was done by compiling a map of each species' present distribution and defining it in terms of temperature variables. The second phase of this exercise involved the selection of dated fossil beetle assemblages from 25 different locations within the British Isles. The mutual climatic range of each of these assemblages was determined by superimposing the climatic range of each species present. A range of possible palaeotemperatures was thereby deduced for each assemblage and the median value calculated. By plotting the age of each assemblage against the deduced palaeotemperature a graph of temperature variation may be obtained as illustrated in the diagram.

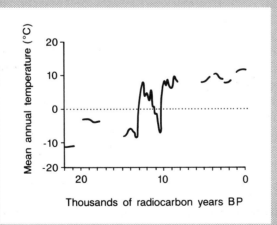

Source: Atkinson, T. C., Briffa, K. R. & Coope, G. R. 1987. Seasonal temperatures in Britain during the past 22, 000 years, reconstructed using beetle remains. *Nature* **325**, 587–592. [Diagram modified from: Atkinson *et al.* (1987) *Nature* **325**, Fig. 3, p. 592]

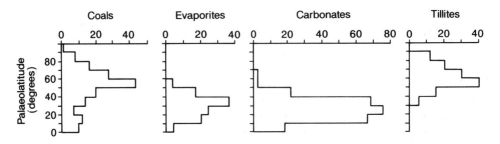

Figure 5.16 *Palaeolatitudinal zonation of climatically sensitive deposits. Each histogram shows the frequency of deposits against palaeolatitude. Evaporites and carbonates are concentrated in the low latitudes, while tillites and coals are mostly found in higher latitudes. [Reprinted by permission from: Scotese & Summerhayes (1986)* Geobyte **1***, Fig. 1, p. 29]*

those indicative of warm or arid climates; (2) those indicative of wet or humid climates; and (3) those which indicate cold or glacial conditions.

1. **Lithologies and deposits indicative of warm or arid climates.** Within the geological record **evaporites** are probably the best indicators of global aridity. Today they are precipitated on or below the surface of the soil in arid or semi-arid regions and include rock salt and gypsum. Evaporites are characteristic of rainfall deficient, subtropical belts with high rates of evaporation. Their occurrence within the geological record provides, therefore, a good indication of arid conditions.

    Sediments with a strong red colouration, which is usually caused by the precipitation of iron oxide, are also sometimes taken as indicative of arid climates, although **red beds** can also occur as a result of fluvial deposition in a variety of environments. The presence of feldspar-rich (arkosic) sandstones has also been used as an indicator of arid climates, but is again somewhat unreliable, since arkosic sandstones can form in a variety of environments.

    When sand dunes, in either coastal or desert regions, become lithified into sandstone a record of the wind regime in which deposition occurred may be recorded. The internal structures of aeolian sandstones can be difficult to interpret but are commonly used to give indications of palaeowind directions (Figure 5.17). Extensive deposits of aeolian sandstone may also be used as indicators of desert-like environments.

    The deposition of calcium carbonate as limestones or chalks may also be used to indicate a warm climate. The solubility of calcium carbonate decreases as temperature increases and therefore high temperatures tend to lead to more limestone. Care is, however, required since the presence of abundant limestone is also correlated with periods of high sea level.

2. **Lithologies indicative of wet or humid climates.** Both bauxite and coal can be used as indicators of wet or humid climates. **Bauxite** is the end product of intense chemical weathering and soil leaching, typical of a tropical location. It develops from weathering of rocks rich in aluminium, at sites which experience intense leaching and where rainfall exceeds 1200 to 1500 mm a year. Its palaeoclimatic

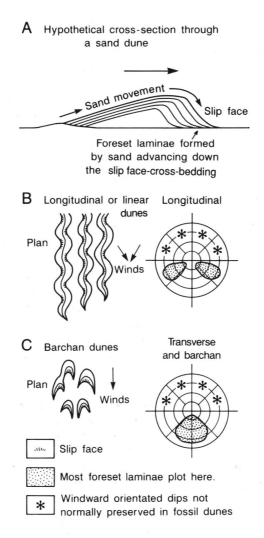

A Hypothetical cross-section through
a sand dune

Sand movement

Slip face

Foreset laminae formed
by sand advancing down
the slip face-cross-bedding

B Longitudinal or linear Longitudinal
dunes

Plan

Winds

C Barchan dunes Transverse
and barchan

Plan

Winds

Slip face

Most foreset laminae plot here.

* Windward orientated dips not
normally preserved in fossil dunes

Figure 5.17 *Dune cross-bedding and palaeowind direction. **A:** shows a hypothetical cross-section through a sand dune illustrating the formation of foreset laminae formed by sand avalanching down the slip face, producing cross-bedding. **B:** shows the pattern of slip faces along a longitudinal or linear (seif) dune. When the orientation and dip of foreset laminae are plotted on a stereogram they plot in the two locations shown (sterogram: orientation is plotted on the radial axis and dip via the concentric lines). **C:** shows the pattern of slip faces on barchan dunes. When the orientation and dip of the foreset laminae are plotted on a stereogram, they plot in the location shown. The diagram illustrates how both palaeowind direction and dune type can be deduced by the analysis of cross-bedding in sand dune deposits. [Modified from: Collinson & Thompson (1989) Sedimentary Structures, Unwin Hyman, Fig. 6.47, p. 90]*

significance lies in its association with high rainfall. The presence of **coal** can also be used to indicate wet or humid conditions. Coal forms from the deposition of vegetation in anoxic conditions where organic material does not break down and decay: today's peat bogs may be tomorrow's coal. Peat bogs generally occur in high latitudes with damp humid climates. Woody material does not normally accumulate in lower latitudes due to the higher temperatures which result in rapid oxidation and breakdown by bacteria. There are however complications, since the principal prerequisite for coal—poor drainage—may occur even in semi-arid areas given a high water-table. Consequently, there are limitations to the use of peat and coal as temperature and precipitation indices, but with care they can sometimes be used.

Fossil soils (**palaeosols**) and weathered layers may also give some indication of palaeoclimate. The characteristics of a soil are partly determined by climate while the intensity of a weathered profile and the products of that weathering are also climatically controlled. Therefore the presence of weathered horizons or soils within the geological record may provide valuable clues as to the former climate.

3. **Lithologies indicative of cold or glacial climates.** There are two main indicators of cold or glacial climates. The first of these is the presence of **tillites**. Glaciers and ice sheets deposit glacial sediments known as tills or boulder clays, which when lithified are known as tillites (Figure 5.18). Tillites provide evidence of glaciation. Modern till usually consists of large subrounded pebbles or boulders set in clay or silt. The pebbles or boulders present tend to be scratched (striated) and aligned in the direction in which the glacier flowed while the sediment was deposited. Caution is, however, required when identifying tillites since deposits produced by mud or debris flows may have similar characteristics. The association of tillites with rock-scoured or striated surfaces is usually considered to be diagnostic of glaciation.

    The second indication of a cold or glacial climate is the presence of **dropstones** in finely bedded (laminated) sediments. Laminated sediments are usually produced by the settling out of sediment from suspension. Periodic variation in the supply of sediment results in the deposition of very fine layers or laminations. Consequently in areas in which laminated sediment is being deposited the only way in which large stones can be deposited without disrupting the laminae is by their vertical deposition from rafts of ice. These rafts may be icebergs or simply sea or lake ice which entrains debris from a shore by freezing. However, dropstones alone cannot be used as evidence of glaciation, because of the variety of other possible rafting agents, such as tree roots.

5.3.4 Reconstructions Based on Computer Climatic Models

Given some basic information about the palaeoclimate of a period further detail and understanding of its climate can be obtained from computer models. There are two main types of climatic model: (1) simple **parametric models**, which examine the interrelationship of a few specific climatic variables; (2) general circulation models which attempt to simulate the whole of the Earth's atmospheric circulation. Models are used not only to predict climatic patterns and behaviour but also to explore the causes of climatic change.

Figure 5.18 *Photograph of glacial till produced by deposition beneath a glacier. Lower photograph shows detail of the till fabric. This till is exposed at Glanllynnau, in North Wales, and dates from the Devensian Glacial. [Photograph: M. R. Bennett]*

5.4 PALAEOECOLOGY

Palaeoecology is the study of the interaction of ancient organisms with their environment and with each other. Study of the palaeoecology of a group of organisms allows deductions to be made about the nature of their environment.

Fossils form a significant component of the sedimentary record. Today, animals and plants inhabit most environments from the polar wastes to the interstices between grains of sand. The relationship of living organisms to each other and to their environment is known as ecology. It is possible with careful observation to deduce the relationship of the once-living and now fossilised organisms within their ancient environment—palaeoecology—by inference from their morphology and by comparison with their living relatives. In essence this is actualism, the same process that we applied to the study of facies. Thus by observing the overall characteristics of a body of rock and by comparing it with the known products of contemporary environments we can make deductions about the nature of ancient environments.

Deductions about palaeoenvironment can be made at two levels, from individual fossils and from the interactions of fossils. Study of individuals helps us to understand their life history better, and it is possible to identify in some well-preserved fossils whether an organism has been attacked by a predator, or whether an organism was suffering from a parasitic infestation or illness. A good example of inference about palaeoenvironment from an individual is that of the preservation of insects in amber, fossil tree sap which not only killed the organism, but also preserved it. Information is thus available about the nature of the forest environment from both the insect and the amber.

Groups of individuals, and particularly groups of individuals of different species, carry the greatest amount of environmental information. The fundamental 'unit' of ecology is the **community**, that is, a group of two or more species which occupy the same habitat. Such communities are controlled today by a series of interacting, environmentally **limiting factors**. Examples of limiting factors include oxygen availability, salinity, substrate, water depth and temperature. The success of organisms and their interaction with each other is based upon their ability to tolerate the limiting factors present. Today, for example, the hermatypic, or reef-building corals, only flourish in reefs where the sea water is: (1) of normal salinity, (2) of shallow depth (<20 m), and (3) with an optimum temperature of 20–29°C (Figure 5.19). As these factors are changed the coral community either flourishes or flounders. Consequently, if one can deduce the limiting factors which controlled a fossil organism or community it is possible to obtain a powerful tool with which to interpret the pattern of changing environments within the stratigraphical record.

There are some important problems associated with the use of fossils in this way. First, it is almost impossible for geologists to be certain whether the assemblage of fossils collected is actually representative of a true community, since most soft-bodied organisms are destroyed before they are fossilised. In addition, fossils from different natural communities may be swept together by action of currents during deposition. There are some exceptions to this; for example in exceptional circumstances a whole living community may be preserved. Such exceptional finds are

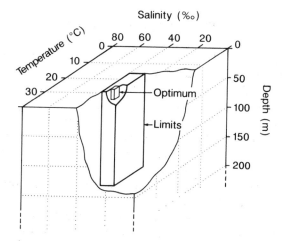

Figure 5.19 *Corals and environmental sensitivity. This diagram illustrates the environmental parameters which limit the growth of present-day reef-building (hermatypic) corals. The optimum growth of hermatypic corals lies within narrow ranges of salinity, temperature and depth. Using uniformitarianism it is possible to predict that fossil corals lived within the same optimum limits, and therefore provide information on the broad palaeoenvironment. [Modified from: Wells (1957) Geological Society of America Memoir 67, Fig. 1, p. 1088]*

known as **Lagerstätten** (fossil bonanzas), but unfortunately these do not occur with suffcent frequency to be of use in the day-to-day interpretation of the stratigraphical record (Box 5.5). Second, with increasing geological antiquity fossil organisms become increasingly different from living organisms and consequently modern analogues cease to be available. For example, the organisms of the Late Proterozoic **Ediacaran Biota** are almost unique and apparently unrelated to most living organisms, and consequently little is known about them.

Despite these problems the fossil record can provide useful information about the nature of palaeoenvironments. For example, by inference from modern descendants, cephalopods, echinoderms and corals are only found in marine environments. By contrast, oysters and mussels are at home in intertidal environments, with uncertain salinities. We can also identify characteristic **biofacies**. Fossil assemblages indicative of oxygen-poor bottom waters or those which typically form reefs are examples of biofacies (Box 5.6). We have already seen that fossils aid us in inferring palaeoclimate, but they can also be of use in determining palaeogeography. For example, the Early Palaeozoic brachiopod assemblages of Wales have been interpreted in terms of water depth and consequently the changing assemblage of brachiopods has enabled shorelines for the Early Palaeozoic to be constructed through time, an example which is described more fully in Section 7.2.3.

Amongst the most important fossils for palaeoecological interpretation of the stratigraphical record are **trace fossils** (Figure 5.20). The study of trace fossils is now commonly called **ichnology**. Trace fossils represent the tracks, trails and feeding traces of living organisms (including droppings, or coprolites). Although it is extremely rare

BOX 5.5: LAGERSTÄTTEN

The fossil record is usually considered to be incomplete. For example, animals with hard parts such as shells and bones are more likely to survive the rigours of the process of fossilisation than those animals that are wholly soft bodied. For the most part, therefore, fossil assemblages are thought to be rarely representative of the once-living communities. Exceptionally, fossils are found in remarkable concentrations (**concentration deposits**) or with extraordinary preservation of soft parts or of soft-bodied organisms (**conservation deposits**). Such finds were referred to as **Fossil Lagerstätten** by the German palaeontologist Adolf Seilacher. **Conservation Lagerstätten** are exceptionally important as they provide rare glimpses of ancient communities and provide a greater understanding of the biology of fossil organisms. Famous Lagerstätten include the Cambrian Burgess Shale in British Columbia, with its exceptional preservation of soft-bodied organisms, the Devonian Ryhnie Chert of Scotland, which preserves an important early terrestrial biota, and the German Solnhofen Limestone, of Jurassic age, which preserves the evidence of the first bird, *Archaeopteryx*. In all of these, preservation has been quick, and decay halted.

Sources: Clarkson, E. N. K. 1993. *Invertebrate Palaeontology and Evolution.* (Third Edition) Chapman & Hall, London. Whittington, H. B. & Conway Morris, S. (Eds) 1985. Extraordinary fossil biotas: their ecological and evolutionary significance. *Philosophical Transactions of the Royal Society*, Series B **311**, 1–192.

for the organism to be found in association with its burrow, or feeding trace, it is clear that there is an association of certain types of trace with certain types of environment (Figure 5.20). Trace fossils are classified on the inferred action of their producer: we can recognise resting traces, movement traces, feeding traces, dwelling traces, and so on. Most importantly, they appear to have a correlation with water depth and consequently depth-related trace fossil biofacies can be recognised (Box 5.7). Trace fossils have a significant advantage over body fossils as they are rarely eroded out of their host rock and reworked into younger deposits, and they are therefore reliable indicators of their environment. Trace fossils are therefore extremely valuable tools in the interpretation of the palaeoenvironment of sedimentary sequences, and often they give a fuller picture of the nature of the biological community than just the shelly fossils alone.

5.5 SUMMARY OF KEY POINTS

- **Sedimentary facies** are defined as the sum total of the characteristics of a body of sedimentary rock, which allow the environment of deposition to be deduced. Those characteristics are commonly the **geometry** of the rock body, its **lithology**, **sedimentary structures** and **fossil content**.

BOX 5.6: OXYGEN-RELATED BIOFACIES

Rhoads and Morse (1971) were first to suggest that the observed sensitivity of organisms to dissolved oxygen could be used to interpret the amount of dissolved oxygen that was present in ancient sedimentary environments. They recognised three oxygen-related biofacies, which they called **anaerobic, dysaerobic** and **aerobic,** based on the levels of dissolved oxygen as illustrated below. In anaerobic facies, bottom-dwelling organisms were absent, consistent with a total lack of oxygen. With dysaerobic facies, some sediment-feeding organisms were seen to rework sediments, although bottom-dwelling organisms were largely absent. This is consistent with very low levels of dissolved oxygen, capable of sustaining very little life. Aerobic facies contain a great variety of marine organisms, both within and on top of the sediment, consistent with increased levels of oxygen. Successful interpretation of black shale sequences using the biofacies concept has enabled geologists to give an estimate of the oxygenation of a basin through geological time.

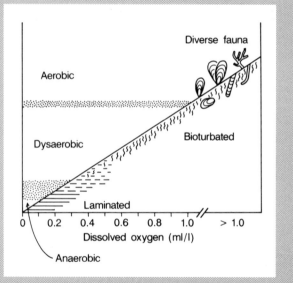

Source: Rhoads, D. C. & Morse, J. W. 1971. Evolutionary and ecologic significance of oxygen-deficient basins. *Lethaia* **4**, 413–428. [Diagram modified from: Rhoads & Morse (1971) *Lethaia* **4**, Fig. 5, p. 421]

- **Facies** are not static: the relative position of facies can change in response to sedimentary or **autocyclic** mechanisms, such as the increased or decreased input of sediment into a sedimentary system. In such systems, the vertical facies variation is equivalent to the lateral facies variation at a specific time (**Walther's principle**).

Figure 5.20 *Photographs of trace fossils. **A:** The upper photograph shows vertical burrows probably made by worms, in Cenozoic sandstones from Norfolk, England. The burrows stand out from the face because they were infilled by coarser grained brown sand and were subsequently cemented by the precipitation of iron through percolating ground water. Hammer for scale [Photograph: BGS]. **B:** The lower photograph shows trails created by organisms, probably arthropods, moving across the surface of the sediment. Coin for scale [Photograph: D. G. Helm]*

BOX 5.7: BATHYMETRY FROM TRACE FOSSILS

It is now commonly accepted that trace fossil assemblages in marine sedimentary rocks are related to the relative depth of accumulation of the sedimentary body. This follows the work of Adolf Seilacher, who recognised that there were a relatively small number of communities of trace fossils which appear limited to different sedimentary facies. Seilacher recognised four distinct assemblages. In coarse, shallow marine sediments the traces were largely simple dwelling traces of worms (e.g. *Skolithos*). Within shelf marine sediments the traces were indicative of a range of crawling (e.g. *Cruziana*), feeding and more complex dwelling traces. Turbidite sequences commonly contained the complex sediment-mining trace *Zoophycos*, while submarine fan sediments were characterised by complex winding, surface-grazing traces (e.g. *Nereites*). Although it is now accepted that the *Skolithos*-type traces are capable of developing in coarse sediments offshore, the *Skolithos, Cruziana, Zoophycos* and *Nereites* trace fossil biofacies (**ichnofacies**) are accepted as representative of progressively more offshore environments, as illustrated. These ichnofacies are therefore powerful tools in palaeoenvironmental analysis.

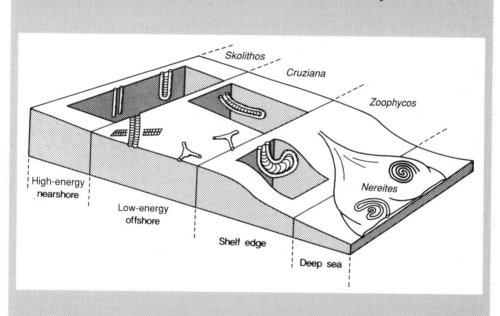

Source: Seilacher, A. 1967. Bathymetry of trace fossils. *Marine Geology* **5**, 413–428. [Diagram modified from: Briggs & Crowther (1990) *Palaeobiology: a Synthesis*, Blackwell, Fig. 2, p. 358]

- External or **allocyclic** influences, in particular **sea level changes**, have a profound influence on the distribution of facies associations. Therefore **transgressive** (onlapping) and **regressive** (offlapping) facies associations may be deduced from the stratigraphical record.
- Relative sea level at the continental margin can be deduced from the interpretation of seismic sections. Sedimentary packages can be recognised on these sections and their relative position on or offshore (their **relative onlap**) can be deduced. This provides a key to relative sea level.
- **'Sequence' stratigraphy** recognises that packages of sediment at the continental margin are commonly bounded by unconformities, which are probably genetically related to sea level changes. Given global sea level changes, 'sequence' stratigraphy is a tool for the correlation of large-scale sedimentary packages.
- Small-scale **palaeogeography** or the ancient geographies of specific time intervals can be interpreted directly from the association of facies in the stratigraphical record.
- **Palaeoclimate** may be deduced directly from the study of **lithology, palaeontology** and **geochemical analysis** of rocks.
- **Palaeoecology** or the ecological relationship of fossils to their ancient environment and to each other can be deduced from the facies relationships of fossils, and is a powerful tool in the reconstruction of ancient environments.

5.6 SUGGESTED READING

Selley (1985) is a nicely written, easily read and very informative book on the nature of sedimentary facies. Duff (1993) gives a good account of sedimentary environments while Collinson and Thompson (1989) is a very important source on the nature and formation of sedimentary structures. The book by Hallam (1992) provides in-depth coverage on sea level changes and provides a useful critique of 'sequence' stratigraphy. Tucker (1988) is a very good handbook providing information on most of the topics in this chapter. Ager (1963) is quite an old text but still important as an introduction to palaeoecological principles. Bromley (1990) provides an elegant introduction to trace fossils. Further information on many of the concepts presented in this chapter is given in Brown *et al.* (1992).

Ager, D. V. 1963. *Principles of Palaeoecology*. McGraw-Hill, New York.

Bromley, R. G. 1990. *Trace Fossils*. Unwin Hyman, London.

Brown, G. C., Hawkesworth, C. J. & Wilson, R. C. L. (Eds) 1992. *Understanding the Earth*. (Second Edition) Cambridge University Press, Cambridge.

Collinson, J. D. & Thompson, D. B. 1989. *Sedimentary Structures*. (Second Edition) Unwin Hyman, London.

Duff, P. McL. D. (Ed.) 1993. *Holmes' Principles of Physical Geology*. (Fourth Edition) Chapman & Hall, London.

Hallam, A. 1992. *Phanerozoic Sea-level Changes*. Columbia University Press, New York.

Selley, R. C. 1985. *Ancient Sedimentary Environments*. (Third Edition) Chapman & Hall, London.

Tucker, M. E. 1988. *The Field Description of Sedimentary Rocks*. Open University Press, Milton Keynes.

6
The Evolution and Closure of Sedimentary Basins: The Role of Plate Tectonics

In the preceding chapters we have introduced the basic stratigraphical tool kit with which the time history of rock units can be determined and each unit interpreted in terms of ancient Earth environments. This chapter now considers the distribution of rocks across the surface of the Earth in the context of the development of the Earth's major structural components and their interaction.

6.1 THE MYSTERY OF MOUNTAIN BUILDING

Early geologists thought in terms of the interpretation of local and regional rock successions. However, throughout the 19th century there was an increasing emphasis on the comparison of the rock succession over very large areas. This was possible with the evolution and increasing sophistication of the stratigraphical tool kit, which has been introduced in previous chapters.

This large-scale or global perspective led to the discovery that many of the thickest sequences of sedimentary rocks tended to be concentrated in long, linear belts. These belts occur along the margins of ancient continental crust known as **cratons** and have been heavily folded and faulted. These belts of deformed rock form the Earth's main tectonic belts (Figure 6.1). They consist of rocks which were deposited in deep sedimentary basins and which have been subsequently folded, faulted, thrust, intruded by igneous rocks and metamorphosed. The uplift of these belts of deformed sediments has produced long chains of mountains which been have carved and shaped to various degrees by erosional processes. The formation of these sedimentary basins and their subsequent compression provided one of the most enduring mysteries for the early

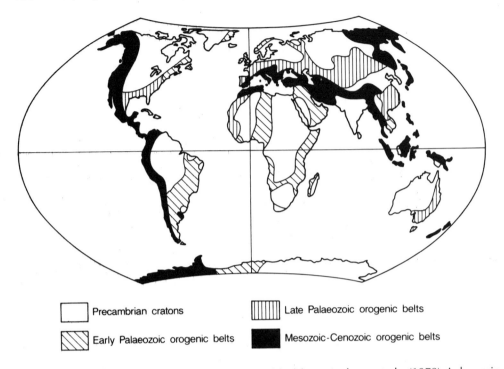

☐	Precambrian cratons	▥	Late Palaeozoic orogenic belts
▨	Early Palaeozoic orogenic belts	■	Mesozoic-Cenozoic orogenic belts

Figure 6.1   *The world's major tectonic belts. [Modified from: Anderton* et al. *(1979) A dynamic stratigraphy of the British Isles,* Allen & Unwin, *Fig. 1.1, p. 3]*

geologist. Ever since James Hutton first discovered the presence of cycles of uplift, erosion and deposition within the stratigraphical record the quest for the mechanism of uplift or mountain building was on.

In the middle of the 19th century geologists began to interpret these tectonic belts in terms of **geosynclines**. Geosynclines were seen as regional troughs or basins which gradually subside as they are filled by sediment. This subsidence led to the intrusion of igneous rocks in the basin. Folding and metamorphism took place when a thick pile of sediment and igneous rocks had accumulated, and this deformation eventually led to the formation of a mountain chain. The mechanism by which deformation was produced was never adequately explained. In the following hundred years the essential elements of the geosynclinal theory were modified and refined. However, it was not until the development of the theory of plate tectonics in the mid-1960s that a coherent picture of the formation of sedimentary basins and their deformation into tectonic belts was obtained. Plate tectonics has unlocked the mystery of mountain building.

6.2 CONTINENTS ADRIFT

The idea that the continents may have moved across the Earth's surface through geological time was explored by several geologists in the 19th and early 20th centuries.

The most convincing evidence was the remarkable fit between the Atlantic coasts of Africa and South America. However, the hypothesis of **continental drift** was largely developed by one man, the German meteorologist Alfred Wegener. Wegener suggested in the early 20th century that the Earth's continents had at one time been joined in two supercontinents. Wegener used as his evidence the fit of the continents already alluded to, but he also utilised many elements of the stratigraphical tool kit. In particular he used two lines of essentially stratigraphical evidence: fossil distribution and lithological similarity.

Wegener studied the distribution of fossil land plants and animals to aid in his interpretations. In particular, he studied the distribution of the plant *Glossopteris*, the leaf remains of which were relatively common in the Southern Hemisphere continents. Wegener reasoned that in order for *Glossopteris* leaves to be found in the widely spaced continents of the Southern Hemisphere then the continents must once have been joined. Using this evidence Wegener grouped all of the southern continents, together with India, into the supercontinent of **Gondwanaland**, now more commonly shortened to **Gondwana**. Wegener also studied the distribution of major geological bodies, such as crystalline basement complexes and mineral deposits. Using these lithological criteria Wegener was able to demonstrate that the fit predicted by map estimates was confirmed by the alignment of geological complexes on either side of the Atlantic Ocean. Wegener presented his findings in his book *Die Entstehung der Kontinente und Ozeane* (The origin of continents and oceans) published in 1915. Although some of Wegener's contemporaries accepted his ideas, these pieces of largely stratigraphical evidence were not considered powerful enough to convince sceptics, who felt that the underlying cause was not clearly explained, let alone possible.

The theory awaited the discovery of **palaeomagnetism** and the development of oceanography. Palaeomagnetism utilises the principle that in molten igneous rocks, or unlithified sediments, any magnetic particles will align themselves with the Earth's magnetic field. This magnetic record is stored within igneous rocks when they cool and within sediments when they become lithified. Deviation in the alignment of such palaeomagnetic particles from the current direction of the Earth's magnetic field shows that the continents have moved. In the 1960s two Cambridge scientists, Fred Vine and Drummond Matthews, discovered that on either side of the Mid-Atlantic Ridge there were a series of linear magnetic anomalies. Strips of ocean crust had alternating magnetic orientations (see Section 6.3). These observations were explained by Harry Hess in terms of a sea floor spreading model by which new oceanic crust forms along mid-ocean ridges as the two halves of an ocean move apart. From these simple observations the theory of plate tectonics developed and is now seen by many as the ultimate control on the stratigraphical record.

6.3 PLATE TECTONICS

The plate tectonic model proposes that the surface of the Earth consists of a series of relatively thin, but rigid, plates which are in constant motion (Figure 6.2). The surface layer of each plate is composed of oceanic crust, continental crust or a combination of both, while the lower part consists of the rigid upper layer of the Earth's mantle

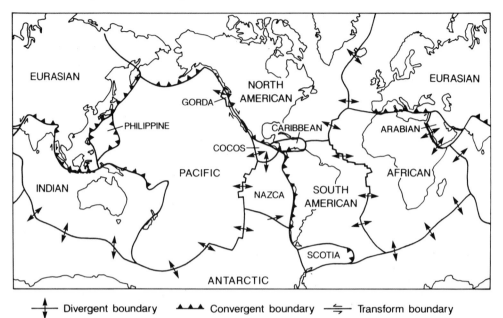

Figure 6.2 *Map of the principal lithospheric plates. The various types of plate boundary are also shown*

(lithospheric mantle). The rigid plates pass gradually downwards into the plastic layer of the mantle, known as the **asthenosphere**. The plates vary in area from $<10^6$ to 10^8 km^2 and may be up to 70 km thick if composed of oceanic crust or 150 km if incorporating continental crust (Figure 6.2). There are seven major plates (10^8 km^2), eight intermediate ones (10^6–10^7 km^2) and over 20 small ones ($<10^6$ km^2). Plates can move at up to 100 mm per year, although the average rate of movement is about 70 mm per year (Box 6.1). Much, but not all, of the Earth's tectonic, volcanic and seismic activity occurs at the boundaries of neighbouring plates. In practice, plate boundaries are recognised by the distribution of volcanoes and earthquakes. There are three types of plate boundaries: (1) **divergent boundaries**, (2) **convergent boundaries** and (3) **transform boundaries**.

1. **Divergent plate margins.** At this type of boundary plates move apart and new oceanic crust is formed in the gap between the two diverging plates (Figure 6.3). Plate area is therefore increased at divergent boundaries. Plate movement takes place laterally away from the plate boundary, which is usually marked by a ridge or rise. This ridge or rise may be offset by **transform faults** (Figure 6.3). Today, most divergent margins occur along the central zone of the world's major ocean basins and the process by which the plates move apart is referred to as **sea floor spreading**. The Mid-Atlantic Ridge and East Pacific Rise provide good examples of this type of plate margin. The rate at which each plate moves apart from a divergent margin varies from less than 50 mm per year to over 90 mm per year and can be determined

BOX 6.1: PLATE TECTONICS: THE FINAL PROOF

In 1984 scientists at NASA released the first direct measurement of plate movements on the surface of the Earth. By measuring the separation of points on the Earth's surface over a period of time it is possible to determine the movement of plates to an accuracy of a few centimetres. These measurements are made with astronomical telescopes and there are about 20 stations around the world which are equipped to make this type of measurement. The measurements are made in one of two ways. In the first method astronomers measure the time taken for two radio telescopes, on different plates, to receive the same signal from outer space. From this it is possible to estimate the rate at which the two telescopes move apart if regular measurements are made. Astronomers have been making measurements of this sort across the Atlantic for over 10 years and have estimated that it widens by about 150 mm every year. The second method involves bouncing laser signals off a satellite equipped with a reflector. The distance of the laser from a satellite is determined by measuring the time taken for the laser beam to be reflected from a satellite and returned to its source. Repeated measurement over a number of years allows the rate at which the plate beneath the laser moves to be measured in relation to the position of the satellite. For example, laser stations have monitored the San Andreas Fault for over 11 years and have found that the two plates which form this transform margin move past each other at a rate of 60 mm every year.

Source: Henbest, N. 1984. Continental drift: the final proof. *New Scientist* **105**, 6.

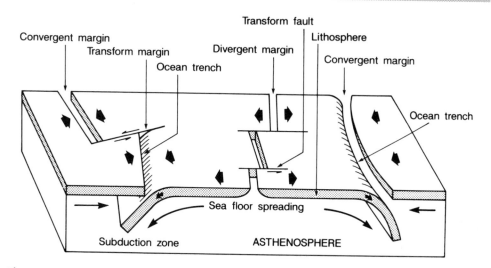

Figure 6.3 *Major plate boundaries. At divergent plate margins plates move apart by sea floor spreading. At convergent plate margins one plate descends beneath the other along a subduction zone and is consumed by the lower mantle. At transform margins plates simply slide past one another. The Earth's lithospheric plates rest on a more mobile layer known as the asthenosphere.*
[Modified from: Isacks et al. (1968) Journal of Geophysical Research **73**, Fig. 1, p. 5857]

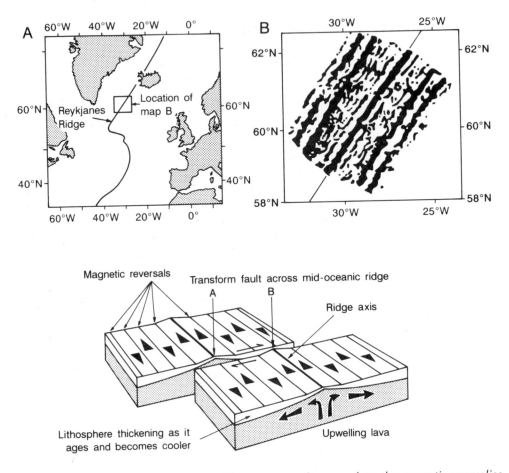

Figure 6.4 *Sea floor spreading centres. The upper two diagrams show the magnetic anomalies across the Reykjanes Ridge in the North Atlantic. The areas of normal polarity are shown in black on diagram B and the ridge axis is indicated by the diagonal line. The lower diagram is a schematic illustration of the process of sea floor spreading and the development of magnetic anomalies. The arrows refer to magnetic polarity. [Map B modified from: Heirtzler et al. (1966),* Deep Sea Research **13**, *Fig. 7, p. 435]*

from the pattern of **magnetic anomalies** either side of a spreading ridge (Figure 6.4). Either side of a spreading centre, weak magnetic anomalies 5–50 km wide and hundreds of kilometres long can be identified. As molten rock cools between diverging plates the magnetic minerals present align themselves with the orientation of the Earth's magnetic field at that time. The polarity of the Earth has changed at regular intervals throughout geological time: magnetic north has alternated between the Arctic (normal polarity) and the Antarctic (reversed polarity) (Table 6.1 and Figure 4.6). Consequently, sections of crust formed during a period of normal polarity have a **palaeomagnetic remnance** which is oriented towards

Table 6.1 *Recent magnetic polarity reversals*

Brunhes	0 to 0.7 million years	Normal polarity
Matuyama	0.7 to 2.4 million years	Reversed polarity
Gauss	2.4 to 3.5 million years	Normal polarity
Gilbert	>3.5 million years	Reversed polarity

today's magnetic north, while a section of crust formed during a period of reversed polarity does not. These long linear strips of magnetic anomalies form a symmetrical pattern either side of a spreading centre (Figure 6.4). A record of the changes in the Earth's magnetic polarity has been established and dated for the Cenozoic and is the basis for magnetostratigraphy (see Table 6.1 and Section 4.2.1). This record, in conjunction with the magnetic stripes found either side of a spreading ridge, allows the rate and pattern of sea floor spreading to be examined.

2. **Convergent plate boundaries.** At a convergent boundary two plates are in relative motion towards each other (Figure 6.3). One of the two plates slides down below the other at an angle of around 45° and is incorporated into the Earth's mantle along a **subduction zone**. The path of this descending plate can be determined from an analysis of deep foci earthquakes and the initial point of descent is marked on the surface by a deep **ocean trench**. Plate area is reduced along the subduction zone. The detailed morphology of convergent margins depends upon whether the converging plates are both composed of oceanic crust or whether one or both of the plates consists of continental crust (Figure 6.5).

   When two plates of oceanic crust collide a **volcanic island arc** may form (Figure 6.5). As one of the plates is subducted beneath the other it begins to melt at a depth of between 90 and 150 km and the resulting magma rises to the surface above the subduction zone to form a chain or arc of volcanoes. The edge of the plate which is not descending is therefore marked by a chain of volcanic islands. If one traversed this type of plate boundary, one would first cross a deep ocean trench and then a volcanic island chain. The distance between the volcanic island arc and the trench is determined by the angle at which subduction occurs. If a plate descends steeply (>45°) then the distance from the trench to the volcanic arc will be small, but if subduction is shallow (<45°) there will be a broader gap between the trench and arc.

   Where a plate composed of oceanic crust meets one of continental crust the former is subducted (Figure 6.5). The molten rock produced in the subduction zone rises to form either a volcanic arc or, alternatively, if the edge of the continental plate is compressed, a chain of volcanoes and mountains are produced along the edge of the continental plate.

3. **Conservative or transform margins.** At a conservative or transform margin two plates move laterally past each other and oceanic crust is neither created nor destroyed. The most famous example of this type of boundary is the San Andreas fault system in southern California.

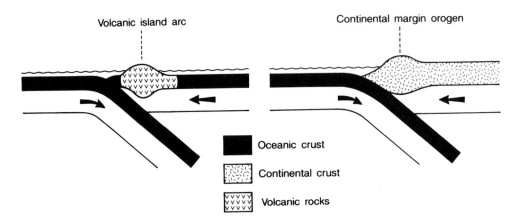

Figure 6.5 *The two main types of convergent plate margin. **A:** This situation occurs where an ocean plate converges with another ocean plate and a volcanic island arc is created. **B:** This situation occurs where an ocean plate converges with a continental plate. Here the ocean plate is subducted beneath the continental margin which may be compressed to form a mountain belt*

There are two fundamental assumptions which underlie the plate tectonic model. First it assumes that the surface area of the Earth has not changed significantly with respect to the rate at which new oceanic crust is generated at divergent margins. The second assumption is that there is little internal deformation within plates relative to the motion between them. If the Earth's surface has increased significantly then sea floor spreading at divergent margins could occur without the destruction of plates at subduction zones; however, most evidence suggests that the Earth's surface area has not changed significantly during the last 500 million years.

In spite of a general acceptance of the plate tectonic model, the question of what causes plates to move has yet to be fully resolved. Four main hypotheses have been put forward to explain this problem. The first hypothesis suggests that flow in the mantle is induced by **convection currents** which drag and move the lithospheric plates above the asthenosphere. Convection currents rise and spread below divergent plate boundaries and converge and descend along convergent margins. The convection currents result from three sources of heat: (1) cooling of the Earth's core, (2) radioactivity within the mantle and crust and (3) cooling of the mantle. The convection hypothesis has been proposed in several different forms throughout the last 60 years but in each case the size of the convection currents and the amount of mantle that is in motion have been different. The second hypothesis invokes the injection of magma at a spreading centre pushing plates apart and thereby causing plate movement. The third possible mechanism involves gravity. Oceanic lithosphere thickens as it moves away from a spreading centre and cools, a configuration which might tend to induce plates to slide, under the force of gravity, from a divergent margin towards a convergent one. The final hypothesis suggests that a cold dense plate descending into the mantle at a subduction zone may pull the rest of the plate with it and thereby cause plate motion. There is still disagreement over the relative merits of these various hypotheses.

The plate tectonic model provides a framework with which to understand the evolution, growth and closure of ocean basins. Ocean basins are formed by the rifting and break-up of continents and grow through the process of sea floor spreading. Within the lifetime of an ocean, subduction will begin and may come to dominate the tectonic processes. As a consequence the ocean will shrink as more sea floor is lost through subduction than is created by sea floor spreading. Eventually the ocean will close and the continents along its margins will collide and fold the sediments which accumulated in the former ocean to form a chain of mountains. The line of closure is known as the **suture**. This complete cycle in the life of an ocean is known as the **Wilson Cycle** and today's oceans can be seen as part of this cycle (Figure 6.6). It is important, however, to note that the Wilson Cycle is essentially a descriptive concept as a given ocean basin may not necessarily follow the full cycle.

In summary, the plate tectonic model provides a mechanism by which: (1) continents can move across the surface of the globe; (2) patterns of volcanism can change and shift across the globe as plates and their boundaries evolve and move; (3) new oceans may grow and different sedimentary basins evolve; and (4) oceans and sedimentary basins close and are deformed to produce mountains. These are mechanisms which are essential to understanding the Earth's stratigraphical record.

6.4 THE EVOLUTION OF SEDIMENTARY BASINS

Most sedimentary rocks are deposited in some form of basin, the size and shape of which is controlled by the processes of plate tectonics. Present-day sedimentary basins provide analogues with which to interpret the depositional environments of sedimentary rocks in the stratigraphical record. The processes which operate at plate margins produce a wide range of different types of sedimentary basins, but in general five broad categories can be recognised (Table 6.2) each of which may be associated with a characteristic sedimentary record (Table 6.3).

1. **Basins and divergent plate margins.** Two types of basin occur in association with plate divergence and therefore with the first half of the Wilson Cycle (Figure 6.6). The formation of a **rift valley** is the first stage in the separation of a piece of continental crust into two plates (Figure 6.7). As the two pieces of continental crust move apart the rift valley increases in width and may subside towards sea level where it may flood to form a shallow sea. Continued plate divergence will widen this seaway and its floor will gradually become composed of oceanic crust. A sea floor spreading ridge will develop along the central axis of the sea and continued divergence will lead to the evolution of an ocean basin (Figure 6.6). The Red Sea is presently at the first phase, produced by the divergence of the Arabian and African plates, while the Atlantic Ocean represents a later stage in the process, since the

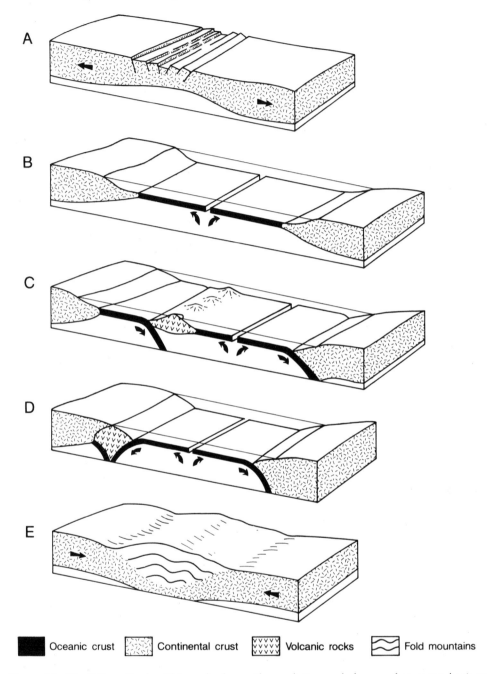

| ■ Oceanic crust | ▨ Continental crust | ⩔⩔ Volcanic rocks | ∿ Fold mountains |

Figure 6.6 *The Wilson Cycle. This cycle shows the evolution and closure of an ocean basin to produce a linear chain of fold mountains. **A:** Rifting of continental crust. **B:** Formation of new ocean crust by sea floor spreading. **C** and **D:** Closure of ocean by subduction. **E:** Following the subduction of all the oceanic crust the two continental plates collide to produce an orogenic belt*

Table 6.2 *Types of sedimentary basin*

1. Basins produced by extension at or close to divergent plate margins
   - Terrestrial rift valleys within continental crust
   - Passive continental margins in mid-plate locations but at the continent–ocean interface
2. Basins formed at convergent plate margins
   - Trenches formed by subduction of oceanic crust
   - Fore-arc basins which develop between ocean trenches and margins of a continental or island arcs
   - Back-arc basins which form behind island arcs
   - Foreland basins formed at continental margins during crustal collision
3. Basins formed along continental transform margins
   - Basins formed along strike-slip fault systems
4. Continental basins unrelated to plate margins
5. Ocean basins

Table 6.3 *Typical sediments within different sedimentary basins. [Based on information in: Dott & Batten (1988) Evolution of the Earth, McGraw-Hill; Nichols (1993) In: Duff (Ed.) Holmes' Principles of Physical Geology, Chapman & Hall]*

Basin type	Characteristic sediment	Depositional environment
Trench	Fine sediment overlying ocean floor basalts	Deep marine
Fore-arc	Heterogeneous gravels, sands and muds derived from erosion of volcanic, metamorphic and granitic rocks of the adjacent volcanic arc or continental margin	Non-marine to marine
Foreland	Heterogeneous gravels, sands and muds derived from the orogenic belt	Mostly river and deltaic
Passive margin	Quartz-rich sands and limestones passing seaward to muds	Shallow marine shelf to deeper marine
Rift valley	Earliest rock volcanic, overlain by thick gravels and sand; younger rocks may include evaporites and limestones	Rivers and lakes changing to shallow marine

North American and Eurasian plates have been moving apart for much longer. It is important to note that the formation of a rift valley does not automatically lead to the rupturing of a plate and the formation of a new ocean basin: the process of divergence may stop if the driving mechanism is shut off.

The trailing edges of the continents which result from the formation of a new ocean are sites of little seismic activity and are not associated with plate movement. They are therefore referred to as **passive margins**: the ocean–continent interface is not marked by a subduction zone and both continent and ocean are part of the same plate. Passive margins are normally associated with large continental shelves and shallow coastal waters, which are ideal for accumulation of a large sediment pile.

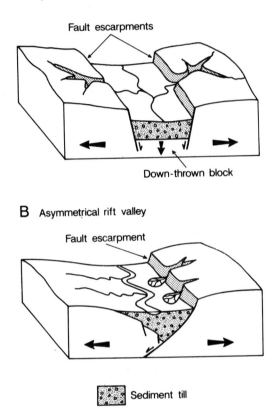

A Symmetrical rift valley

Fault escarpments

Down-thrown block

B Asymmetrical rift valley

Fault escarpment

Sediment till

Figure 6.7 *The structure and morphology of rift valleys. **A:** A symmetrical rift valley, or graben.*
***B:** An asymmetrical rift valley, or half-graben. [Modified from: Summerfield (1991)* Global
Geomorphology, *Longman, Fig. 4.9, p. 92]*

Such sediments tend to spread from these shelf environments into the deeper ocean
basins via large subaqueous flows known as turbidity currents. The Atlantic
seaboards of North America and Europe provide good examples of passive margins.

2. **Basins and convergent plate margins.** Four types of basin are associated with
convergent plate margins and are generally related to the second half of the Wilson
Cycle (Figure 6.6). The destruction of oceanic crust occurs along subduction zones,
which are marked on the surface of the ocean floor by a deep trench (Figures 6.3
and 6.8). **Ocean trenches** are typically about 2 km deeper than the level of the ocean
floor and are often more than 6 km below sea level. Trenches form the deepest parts
of ocean basins; for example the Marianas Trench in the Pacific Ocean is more than
10 km below sea level. Ocean trenches act as large traps for sediment produced
either on volcanic island arcs or from continental margins. As a plate is subducted,

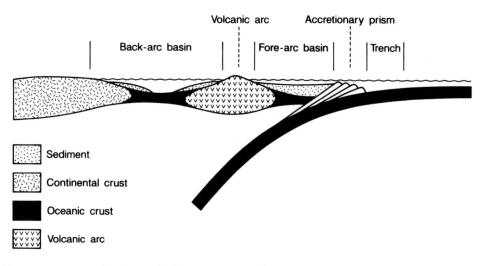

Figure 6.8 *Types of sedimentary basin associated with a subduction zone. Not all these basins may be present along any one margin*

trench sediment may either be taken down into the mantle or alternatively it may scraped off by the overriding plate and added to it in a series of slices. The repeated addition of slices at the edge of the overriding plate may build a triangular body of sediment along the outer edge of a trench known as an **accretionary prism** (Figures 6.8 and 6.9). This triangular body of sediment may define a basin between the trench and the continental margin or island arc, known as a **fore-arc basin** (Figure 6.8). The width of a fore-arc basin is controlled by the distance between the accretionary prism at the trench and the continental margin or island arc, which is determined by the angle of subduction. If the descending plate plunges steeply into the subduction zone then the fore-arc basin will be narrow and conversely if the angle of subduction is shallow then the fore-arc basin will be wider.

A **back-arc basin** may form to the rear or landward side of an island arc (Figure 6.8). The geometry of back-arc basins can be quite complex and their size usually depends upon the position of the island arc in relation to other arcs or to the continental margin.

Subduction of crust in a trench can eventually consume an ocean. As the ocean closes it brings two continents together to the point where they collide in the final stage of the Wilson Cycle (Figure 6.6). The collision of these two continents causes intense deformation of the sediments deposited in the former ocean. The deformed sediments form large thrusts and huge folded sheets of rock called **nappes** which move both upwards and laterally away from the point of collision. These thrust sheets load the crust adjacent to the area of collision causing it to subside under their weight. This loading may form a peripheral basin known as a **foreland basin** which quickly fills with sediment as the thrust sheets and nappes forming a new range of mountains erode.

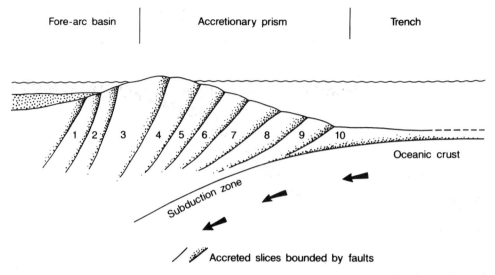

Figure 6.9 *A model of an accretionary prism. As the ocean crust descends slices of trench sediment are accreted to the opposite side of the trench. Each slice is usually fault bounded and the numbers refer to the age of each slice, 1 is the oldest and 10 is the youngest. [Modified from: Nichols (1993). In: Duff (Ed.)* Holmes' Principles of Physical Geology, *Chapman & Hall, Fig. 30.14, p. 710]*

3. **Basins formed at transform margins.** At transform plate margins crust is neither created nor destroyed. However, movement between the two plates is not smooth and if the plate edges are irregular, zones of local compression (**transpression**) and extension (**transtension**) are created as the plates move past one another. Basins formed by transtension are characteristically deep in relation to their size and their margins are associated with steep faults (Figure 6.10). Examples of this type of basin include the Dead Sea rift and the Gulf of California. It is possible to identify a cyclic evolutionary history for this type of basin similar to that recognised for ocean basins by Wilson. The first phase of the cycle involves transtension and the formation of a basin which quickly fills with both coarse alluvial gravel and fine-grained marine or lacustrine sediments. Continued extension and basin subsidence leads to the introduction of volcanic rocks in the floor of the basin. In the second phase of the cycle transpression causes the closure and deformation of the basin.

4. **Continental basins unrelated to plate margins.** A variety of different types of sedimentary basin may form within a plate composed of continental crust. Of these basins the **intercratonic basin** is perhaps the most noteworthy. This type of basin is a broad regional downwarp in continental crust, well removed from the tectonic influence of a plate margin. Such basins do not usually involve faulting, although some, but not all of the basins may be centred on local rift valleys. They are commonly oval in shape and are enclosed, with a central point much lower than the surrounding area. The processes by which these broad downwarps form is not

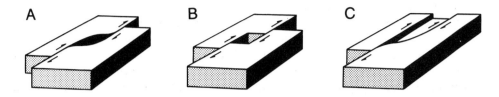

Figure 6.10 *Basins formed along transform margins or strike-slip faults. **A:** A 'releasing bend' in a single fault creates a gap when the fault moves. **B:** an offset in a fault produces a 'pull-apart' basin. **C:** A branch in a strike-slip fault produces a region of extension between two faults. [Modified from: Reading (1980). In: Ballance & Reading (Eds) Sedimentation in Oblique-slip Mobile Zones, Special Publication of the International Association of Sedimentologists No. 4, Fig. 3, p. 12]*

entirely clear, although it has been suggested that they result from the activity of thermal plumes in the underlying mantle. Most volcanic activity is confined to plate boundaries, but local **hot spots** or **thermal plumes** do occur within the interior of plates. If a thermal plume rises beneath a continent it will cause the overlying crust to thin and stretch to produce a rift valley. If this process continues the crust will separate to form two new plates and a new ocean. However, if the thermal event comes to an end before crustal separation occurs then a broad crustal down-sagging may result when the heat is removed, which will give rise to an intercratonic basin. Alternatively thermal plumes may elevate, but not cause rifting. At the end of the thermal episode the crust will subside back to its original position. If it has, however, been thinned by erosion while elevated the crust will subside below its original level to form a basin. In this way regional downwarps may form which are not centred on local rift valleys.

5. **Ocean basins.** Ocean basins are the largest type of sedimentary basin, although the rate of sedimentation in them is often very low due to the distance of the central parts of the basin from land and a major source of sediment.

In conclusion, it is important to emphasise that plate tectonics is a dynamic process and that sedimentary basins may evolve from one type or setting to another through time. For example, as the Wilson Cycle develops a rift valley evolves into an ocean basin. The margins of the ocean basin may then become back-arc or fore-arc basins as subduction of ocean floor commences once it is mature. In referring to a sedimentary basin we are therefore not considering a constant feature but one which changes in character through time.

6.5 THE CLOSURE OF SEDIMENTARY BASINS: THE PROCESSES OF MOUNTAIN BUILDING

Mountain building (**orogenesis**) involves the compression and deformation of a sedimentary basin. There are four types of mountain building processes: (1) **continental margin orogenesis**, (2) **continental collision**, (3) **collage collision** and (4) **transpression**.

1. **Continental margin orogenesis.** Not all mountain chains form by the closure of ocean basins and orogenesis may occur along plate margins which are not undergoing collision (Figure 6.5). This type of mountain building is referred to as continental margin orogenesis and involves the production of a mountain chain along a convergent margin where an oceanic plate is descending beneath a plate composed of continental crust. Compression and deformation will occur at the plate margin if the rate at which plate convergence is occurring exceeds the rate of subduction. The subduction rate is controlled by the rate and angle at which the plate of ocean crust descends beneath the plate of continental crust. This is controlled by the buoyancy of the ocean plate or its temperature—a cold plate for example will descend easily. This in turn is determined by the distance between the centre of sea floor spreading and the subduction zone: the greater the distance the greater the opportunity for the ocean crust to have cooled. If compression occurs it leads to buckling and deformation of the leading edge of the continental plate which is deformed into a chain of mountains parallel to the subduction zone. The Andes provide an excellent example of this type of orogen. Mountain building here is caused by the buckling of the front edge of the continental plate and by the intrusion of large volumes of molten igneous rock which rise from the subduction zone.

2. **Continental collisions.** This involves the production of a range of mountains by the closure of an ocean basin and the collision of two continents along its margin (Figure 6.11). It is this type of orogenic event which is represented by the Wilson Cycle (Figure 6.6). The Himalayas provide the best contemporary example of this sort of orogenic episode. Over the last 80 million years or so the Indian subcontinent has moved progressively northwards as part of the Indian plate. The ocean between the subcontinent and the Eurasian landmass has slowly closed leading to collision between the Indian subcontinent and Eurasia. The Himalayas were produced by this collision.

3. **Collage collisions.** Mountain building may occur as a result of the collision of several different plate elements or fragments of ocean floor. At its simplest, island arcs may collide with continents which in turn may collide with other continents and in this way build up a chain of mountains in successive stages by the addition of successive distinct elements (Figure 6.11). It is probable that any structures which rise above the general level of the ocean floor, such as seamounts, oceanic islands, basaltic plateaux, island arcs or continental fragments, will be accreted to a continental margin when they reach a subduction zone, rather than being subducted. In this way a variety of different structures may become jammed together to form part of a mountain belt.

   Small slivers of continental crust, known as micro-continents, are commonly involved with this type of collision. Each island arc or micro-continent that is added to the collision is called a **terrane** and each terrane may have its own stratigraphical record and different fossil assemblages (Box 6.2). Terranes may not only consist of micro-continents and island arcs but also a wide variety of other types of topographic fragments, such as seamounts, ocean ridges, or accretionary prisms. This type of mountain building is known as **collage tectonics** because of the wide variety of different elements (the terranes) which are involved.

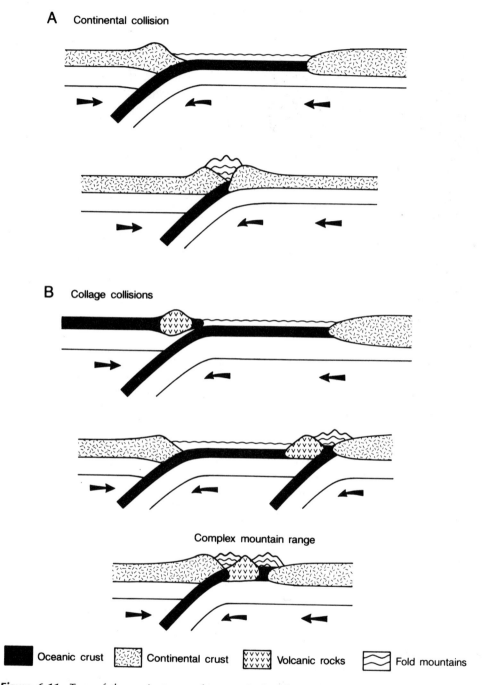

A Continental collision

B Collage collisions

Complex mountain range

■ Oceanic crust   ▨ Continental crust   ▼ Volcanic rocks   〰 Fold mountains

***Figure* 6.11** *Two of the main types of mountain building (orogenesis).* **A:** *Collision of two continents.* **B:** *An example of a collage collision involving two continents and an island arc*

BOX 6.2: TERRANES

A **terrane** is a mappable structural entity which has a stratigraphical sequence and an igneous, metamorphic and structural history quite distinct from those of adjacent units. Terranes are separated from each other by a distinct structural break. They are commonly referred to as **allochthonous,** which means that the associations of rocks that comprise each terrane were not formed in their present position, and that they have been moved. Many modern and ancient mountain chains appear to consist of a sequence of terranes which have been accreted to a continental margin at a subduction zone to form part of either a continental orogen or the core of an orogenic belt produced by the collision of two continents.

The Western Cordillera of North America provides an excellent example of a contemporary continental orogen formed by the accretion of terranes. Over 50 major terranes have been identified within the Western Cordillera, formed by the accretion of a wide variety of rock bodies to the North American continent, such as seamounts, oceanic islands, basaltic plateaux and island arcs (see Section 7.3.1).

Source: Barber, T. 1985. A new concept of mountain building. *Geology Today* **14,** 116–121.

There is increasing evidence that many ancient orogenic belts contain a complex collage of tectonic terranes consisting of micro-continents, island arcs and fragments of ocean floor all jammed together. Recognition of the boundaries between these different terranes, dating their respective collisions and determining the origin of each terrane, is one of the most challenging areas of current research (see Section 7.3.1).

4. **Transpression**. This involves deformation along a transform margin and the closure of any transform basins present (Figure 6.10). Transpression forms an important part of many orogenic episodes since there is usually a strong lateral component within most types of continental collision (see Section 11.4.3).

6.6 SUMMARY OF KEY POINTS

- The Earth's surface consists of seven large plates, eight intermediate ones and over 20 small ones.
- There are three types of plate boundary: (1) **divergent boundaries** or spreading centres, (2) **convergent boundaries** and (3) **transform** or **conservative boundaries**.
- Plate tectonics explains the formation and geometry of many of the Earth's sedimentary basins today.
- The closure of sedimentary basins usually results in the formation of a mountain belt (orogenesis). There are four main types of orogenesis: (1) continental margin orogenesis, (2) continental collision, (3) collage collison and (4) transpression.

6.7 SUGGESTED READING

There are many good accounts of plate tectonics in the literature, but both Dott & Batten (1988) and Duff (1993) provide particularly good summaries. Cox & Harte (1986) provide an excellent, although detailed account of plate tectonics. The chapter on plate tectonics in Hallam (1990) provides an interesting historical background to the development of the plate tectonic model. This is complemented by good chapters on the development of the concept of plate tectonics and of the principles in Brown *et al.* (1992). Barber (1985) provides an excellent account of the process of mountain building in the context of terranes.

Barber, T. 1985. A new concept of mountain building. *Geology Today* **14**, 116–121.
Brown, G. C., Hawkesworth, C. J. & Wilson, R. C. L. (Eds) 1992. *Understanding the Earth.* Cambridge University Press, Cambridge.
Cox, A. & Harte, R. B. 1986. *Plate Tectonics: How it Works.* Blackwell, Oxford.
Dott, R. H. & Batten, R. L. 1988. *Evolution of the Earth.* (Second Edition) McGraw-Hill, New York.
Duff, P. McL. D. (Ed.) 1993. *Holmes' Principles of Physical Geology.* (Fourth Edition) Chapman & Hall, London.
Hallam, A. 1990. *Great Geological Controversies.* (Second Edition) Oxford University Press, Oxford.

7
The Stratigraphical Tool Kit: Case Studies

In this chapter we provide a series of extended examples or case studies which illustrate the application of the stratigraphical tool kit introduced in the previous chapters. The intention is not to illustrate every principle but simply to give an insight into some of the methods by which stratigraphers build up a picture of events in Earth history. For convenience our examples have been grouped under three headings: (1) establishing and dating the sequence; (2) interpretation of the sequence; and (3) the application of stratigraphy to plate tectonics.

7.1 ESTABLISHING AND DATING THE ROCK SEQUENCE

Within this section there are two case studies. The first examines the work which unravelled the chronology of the Lewisian rocks of the Scottish Precambrian. It demonstrates the importance of cross-cutting relationships in determining the relative order of rock units within an area. The second case study illustrates some of the concepts of 'sequence' stratigraphy and discusses its application in field studies.

7.1.1 Unravelling Precambrian Chronology
(Sutton & Watson 1951; Giletti *et al.* 1961)

1. **Introduction**
   The metamorphic Lewisian complex of the Scottish Highlands contains some of the oldest rocks in Britain, dating from the Archaean (Figure 7.1). The Lewisian forms the basement for much of the younger rock cover in Scotland. The complex

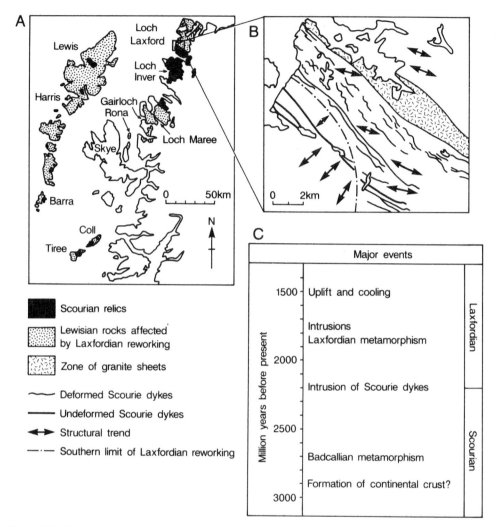

Figure 7.1 *The Lewisian of north-west Scotland. **A:** Distribution of Laxfordian reworking in the Lewisian rocks. **B:** Map of the Scourian to Laxfordian boundary in the Loch Laxford area, showing the contrast between the deformed and undeformed Scourie dykes. **C:** Chronology of Lewisian events. [Modified from: Anderton et al. (1979) A Dynamic Stratigraphy of the British Isles, Allen & Unwin, Figs 2.2 & 2.3, pp. 16 & 17]*

has been subject to several episodes of intense metamorphism and as a consequence developing a stratigraphy for it has proved difficult.

The Lewisian is composed of gneisses which vary in composition from ultrabasic (low silica content) to acidic (high silica content). Early investigations of the complex were undertaken by members of the Geological Survey of Great Britain in the early part of the 20th century. They established a basic chronological framework. These gneisses were observed to have been intruded by a younger group of basic igneous intrusions (dykes) which were later metamorphosed into basic gneisses. This relative sequence of rocks and events within the Lewisian was not challenged until the work of Sutton and Watson (1951). Through detailed field observations and the interpretation of simple cross-cutting relationships Sutton & Watson (1951) were able to show that two phases of metamorphism had occurred: one before the intrusion of the dykes and one after. Sutton & Watson (1951) related their metamorphic episodes directly to orogenic events and suggested that the igneous intrusions represented a punctuation mark between them (Figure 7.1). However, what Sutton and Watson lacked was an independent means of confirming the time equivalence of the igneous intrusions: did they all form at the same time? They also had no clear idea of the time scale over which these orogenic episodes took place. This was provided 10 years later by the application of radiometric dating. Giletti *et al.* (1961) represents one of the earliest attempts at applying radiometric dating techniques to interpret complex field relationships. Giletti and his colleagues were able to put firm constraints on the timing of the metamorphic events.

2. **Cross-cutting relationships—Sutton & Watson (1951)**
    From a basis of field observation in the Loch Torridon and Scourie areas of the North West Highlands of Scotland, Sutton and Watson (1951) were able to subdivide the Lewisian gneisses on the basis of lithology. The relative chronology of these lithological units was established on the basis of the intrusion, tectonic and metamorphic history of a series of north-west to south-east trending dolerite dykes, known as the Scourie dykes. They recognised that the gneisses exposed between the Gruinard Bay and Scourie were of high metamorphic grade (granulite facies) and possessed folded structures which had a north-east trend. Sutton and Watson demonstrated that these rocks were traversed by a swarm of north-west to south-east trending dolerite dykes which were unaffected by this metamorphism (Figure 7.1). The dykes displayed chilled contacts with the gneisses. This clearly illustrated that the dykes were intruded into the gneiss long after it had cooled. Clearly, therefore, the intrusion of the dykes postdated the metamorphism of these gneisses. Farther to the north and to the south of this area, the gneisses were of different composition, rich in the mineral amphibole, typical of lower grade metamorphism. The structural grain in these rocks trended to the north-west (Fig. 7.1). These rocks had also been intruded by the dolerite dykes, but in this case the dykes had also suffered metamorphism and deformation (Figures 7.1 and 7.2). On this basis, Sutton and Watson (1951) developed their model of two major episodes of metamorphism in the development of the Lewisian complex: an older phase, the Scourian, and a younger phase, the Laxfordian, which reworked the Scourian gneisses and the intervening dykes. They regarded the Scourie dykes as time significant, representing an event between the two episodes of metamorphism: 'the period of dolerite intrusion may therefore be taken as a fixed point in the history of the North West Highlands' (Sutton and Watson 1951, p. 292).

3. **Unravelling the ages—Giletti *et al.* (1961)**
    Giletti *et al.* (1961) were able to place firm constraints on the timing of the metamorphic episodes recognised by Sutton and Watson (1951) in the field. They utilised

Figure 7.2 *Photograph of Scourie dykes deformed during the Laxfordian orogeny, from the Isle of Barra, north-west Scotland. [Photograph: BGS]*

rubidium–strontium dating of micas and potassium feldspars in the gneisses. The metamorphism of the Scourian complex was dated to between 2700 and 2200 million years ago. The Laxfordian was dated in the range of 1850 to 1500 million years ago. Giletti *et al.* confirmed that these dates did not indicate one long extended period of metamophism but rather two separate geological episodes between which there was a period of dyke intrusion (Figure 7.1). This work supported and enhanced the original conclusions of Sutton and Watson.

4. **Comment**

This case study demonstrates the importance of cross-cutting relationships in determining the relative chronology of events within an area. In particular it illustrates the application of these tools to structurally complex, high-grade metamorphic terrains and shows that the application of simple stratigraphical techniques can provide important information in determining the relative sequence of geological events within an area. Further information about this story can be obtained in the excellent review by Rogers and Pankhurst (1993).

5. **References**

Giletti, B. J., Moorbath, S. & Lambert, R. St. J. 1961. A geochronological study of the metamorphic complexes of the Scottish Highlands. *Quarterly Journal of the Geological Society of London* **117**, 233–272.

Rogers, G. & Pankhurst, R. J. 1993. Unravelling dates through the ages: geochronology of the Scottish metamorphic complexes. *Journal of the Geological Society of London* **150**, 447–464.

Sutton, J. & Watson, J. 1951. The pre-Torridonian metamorphic history of the Loch Torridon and Scourie areas in the north-west Highlands and its bearing on the chronological classification of the Lewisian. *Quarterly Journal of the Geological Society of London* **56**, 241–295.

7.1.2 The Application of 'Sequence' Stratigraphy in Field Studies: An Example from the Early Palaeozoic Welsh Basin
(Woodcock 1990)

1. **Introduction**

    The concept of 'sequence' stratigraphy was developed for use with subsurface borehole and seismic data. However, the concept has recently been applied to onshore basins using outcrop data, and it has been found to provide a valuable tool in determining the regional history of a sedimentary basin. A good example of this type of application is provided by Woodcock's (1990) work on the Early Palaeozoic Welsh Basin. During the Palaeozoic much of the area which we now know as Wales was a deep sedimentary basin along the southern margin of an ocean called the Iapetus Ocean. In his paper Woodcock erected a new stratigraphy for this basin based on the principles of 'sequence' stratigraphy and using this he established a chronostratigraphical division of these rocks.

2. **Relevant principles of 'sequence' stratigraphy**

    As we discussed in Section 5.1.2, a depositional 'sequence' is a three-dimensional rock body, within a sedimentary basin. The boundaries of a 'sequence' are defined for the most part by unconformities (chronostratigraphical breaks). It must, however, be remembered that individual 'sequence' boundaries may grade from a clear unconformity at the margin of a basin, through disconformity, to conformity at the basin centre (Figure 5.9). The presence of basin-wide unconformities can be used to group 'sequences' into **megasequences**.

    A tentative guide to the 'sequence' stratigraphy of a basin can be obtained by calculating a **rock preservation curve**. A rock preservation curve for an area shows the amount of rock preserved at successive times as a percentage of the potential maximum. The curve is computed from stratigraphical logs, derived from boreholes or rock sections, and the method by which it is calculated is shown in Figure 7.3. On a rock preservation curve megasequence boundaries are recorded as 0% lows on the curve, representing zero preservation across the basin (i.e. no rock deposition anywhere in the basin). The remaining lows represent unconformities which are less than basin-wide and are therefore 'sequence' boundaries.

3. **Application to the Welsh Basin**

    Woodcock (1990) drew up a rock preservation curve for the Welsh Basin on the basis of numerous stratigraphical logs (Figure 7.4). On this curve he was able to recognise four points at which there was zero preservation, which he used to define three megasequences. He then established which lithostratigraphical units belonged to which megasequence by reference to the original stratigraphical logs. On

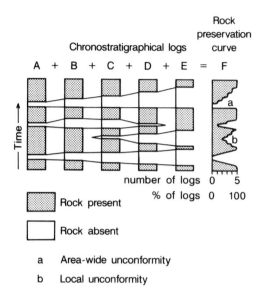

Figure 7.3 *Construction of a rock preservation curve (F) by totalling the amount of rock present in a series of chronostratigraphical logs (A–E) located throughout a depositional basin. [From: Woodcock (1990)* Journal of the Geological Society of London **147**, *Fig. 2, p. 538. Reproduced by permission of the Geological Society of London]*

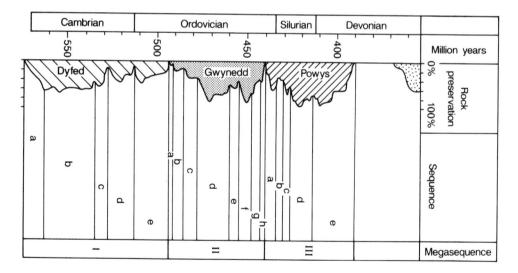

Figure 7.4 *Rock preservation curve for the Early Palaeozoic Welsh Basin showing the main sequences and megasequences identified (% rock volume through time). [From: Woodcock (1990)* Journal of the Geological Society of London **147**, *Fig. 3b, p. 539. Reproduced by permission of the Geological Society of London]*

this basis he was able to set up three supergroups—the Dyfed, Gwynedd and Powys supergroups—corresponding to the three megasequences present. A further 19 'sequence' boundaries were also identified within the curve, giving 18 distinct depositional sequences. In this way he was able to establish a basin-wide stratigraphy for the Welsh Basin, in which individual records or sections could be placed.

Woodcock went on to interpret the 'sequence' stratigraphy he had established. He recognised that the majority of the proposed 'sequence' boundaries in the Welsh Basin were probably caused by processes other than just purely eustatic sea level change. This conclusion is of considerable importance since traditionally 'sequence' stratigraphies are simply interpreted in terms of eustatic sea level variation. Woodcock (1990) and others have suggested that 'sequence' stratigraphies are controlled by tectonism, sedimentation and eustatic sea level variation. Consequently, sea level curves deduced from 'sequence' stratigraphy may not necessarily give a precise picture of eustatic sea level.

4. **Comment**

This case study illustrates two processes of investigation. First, detailed analysis of the stratigraphical record to identify stratigraphical breaks. Second, the application of these data in order to construct a chronostratigraphy, which provides the temporal framework with which to group lithostratigraphical units.

5. **Reference**

Woodcock, N. H. 1990. Sequence stratigraphy of the Palaeozoic Welsh Basin. *Journal of the Geological Society of London* **147**, 537–547.

7.2 INTERPRETING THE ROCK SEQUENCE

Within this section there are two case studies which examine how rock units can be interpreted in terms of palaeogeography, palaeoclimate and palaeoecology. The first case study deals with a facies interpretation of the palaeoenvironments represented by the Permian rocks of the North Sea. The second case study deals with the application of fossil assemblages (biofacies) in determining palaeoshorelines.

7.2.1 Permian Redbeds of North-west Europe: An Example of Palaeoenvironmental Reconstruction
(Glennie 1972)

1. **Introduction**

The Permian redbeds of north-west Europe, known as the **Rotliegendes**, are clastic sediments originally deposited on land, in arid or semi-arid conditions. Their depositional environment is primarily a function of the development of a semi-arid environment which resulted from the construction of a large supercontinent at the end of the Palaeozoic (Pangaea). These sediments occur extensively beneath the North Sea and are host to much of the natural gas present in this area. As a consequence they have been the subject of considerable research. Glennie (1972) provided a remarkable survey of the facies present within the Rotliegendes and his contribution is considered by many to be a classic example of facies analysis. His approach was first, to review the sediments found in modern desert environments and second, to use these modern analogues to interpret the sediment cores taken from boreholes drilled beneath the North Sea.

2. **Sediments of modern deserts**
   Glennie (1972) identified two main types of desert sediments, those deposited by .water and those deposited by wind.
   a. Water-lain sediments—Glennie recognised that there are two main types of water-lain sediments present in desert environments: distinct fluvial sediments and temporary desert lake sediments.
       i. Desert fluvial sediments. Although they are often infrequent, heavy rain-storms do occur in deserts and, because of the absence of vegetation, they are able to move large amounts of debris and cut valleys known as **wadis**. The ephemeral rivers produced by these storms often dry out before reaching the sea or inland lake and therefore dump their sediment load as an unsorted conglomerate within their wadis.
       ii. Temporary desert lakes and inland **sabkhas**. Temporary lakes may form in the centre of a basin with inland drainage, or where sand dunes block a wadi channel. As the water evaporates, salts are deposited and may form a crust of halite or gypsum salt over the sediment surface. Flat areas of salt-encrusted sand, silt and clay are known as sabkhas. Another feature common in this type of environment is the **sand dyke**. A layer of silt and clay is commonly deposited on the floor of a temporary lake as it dries out. This clay layer will become desiccated as the lake dries up and cracks will form. These cracks frequently become filled with a slurry of wet sand from below the dry surface and the structures produced are known as sand dykes. Alternatively, the desiccation cracks may become filled by wind-blown sand from above. **Adhesion ripples** may also form in this type of environment. When dry sand is blown across a wet sediment surface, some grains stick to the surface on impact. If this process continues, an irregular, blistered or ripple-like surface may develop.
   b. Wind deposits (aeolian)—Glennie recognised that wind-blown sand is deposited either in continuous sheets or as discrete dunes (Figure 7.5). Internally, sand dunes contain distinct sedimentary structures (cross-bedding) which not only allow the type of dune to be inferred but also provide evidence of the palaeo-wind direction (Figure 5.17). Aeolian sediments may occur in association with water-lain sediments. The fluvial sands of ephemeral rivers form an important source of wind-blown sand and during the long intervals between rainstorms when the wadi channels are dry the fluvial sands may be reworked by the wind into small dunes. When the wadi is next in flood these aeolian sands are often removed or 'flushed out'.

3. **The sediments of the Upper Rotliegendes of the North Sea**
   Figure 7.6 presents two generalised sedimentary sections taken from the Upper Rotliegendes rocks beneath the North Sea. They are based on numerous sediment cores drilled during the exploration for natural gas and oil. Glennie (1972) inter-preted the sedimentary structures present within these terrestrial redbeds on the

Figure 7.5   *Photograph of a linear sand dune from the Namibian desert. [Photograph: J .E. Bullard]*

basis of the modern desert sediments reviewed in the first part of his paper. From an analysis of the cross-bedding present within the aeolian sands, Glennie (1972) was also able to conclude that both barchan and seif dunes were present and that the palaeowind blew predominantly from east to west.

By looking at the spatial distribution of sediment facies beneath the North Sea in sediment cores, Glennie (1972) was able to produce a conceptual cross-section or model of the facies distribution within the Upper Rotliegendes rocks beneath the North Sea and the Netherlands (Figure 7.7). In the south, this cross-section shows wadi sediments spreading out from the Hercynian mountains. These sediments pass into an extensive dune field beyond which there was a broad desert lake subject to periodic desiccation in the north.

4. **Comments**

The work of Glennie (1972) provides a classic example of the use of facies analysis to give a picture of an ancient period in Earth history. The application of modern analogues emphasises the importance of uniformitarianism in facies analysis. It also shows how a palaeogeographical model of the distribution of the desert environments may be drawn up from borehole and outcrop data. This story has recently been reviewed and updated by George and Berry (1993).

5. **References**

George, G. T. & Berry, J. K. 1993. A new lithostratigraphy and depositional model for the Upper Rotliegend of the UK Sector of the Southern North Sea. In: North, C. P. & Prosser,

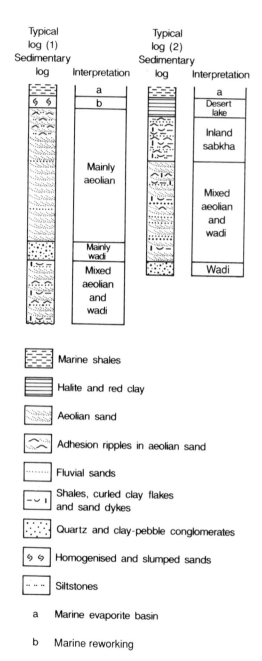

Figure 7.6 *Two idealised sedimentary logs from boreholes through the Rotliegendes of the North Sea. The facies interpretation is shown to the right of each log. [Modified from: Glennie (1972) Bulletin of the American Association of Petroleum Geologists 56, Fig. 9, p. 1055]*

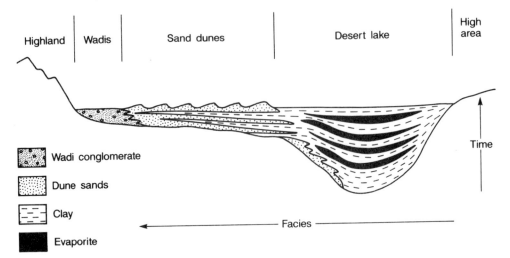

Figure 7.7 *Environments and facies associations of the Rotliegendes of the North Sea, along a transect running north from the Hercynian mountains located in the southern North Sea. [Modified from: Glennie (1972) Bulletin of the American Association of Petroleum Geologists **56**, Fig. 17, p. 1067]*

D. J. (Eds) *Characterization of Fluvial and Aeolian Reservoirs.* Geological Society Special Publication No. 73, 291–319.

Glennie, K. W. 1972. Permian Rotliegendes of northwest Europe interpreted in the light of modern desert sedimentation studies. *Bulletin of the American Association of Petroleum Geologists* **56**, 1047–1071.

7.2.2 Silurian Coastlines
(Ziegler 1965)

1. **Introduction**
    Living organisms can be grouped into communities of two or more species which occupy the same environment or 'ecospace'. The balance of different organisms in a given environment depends on their relative success, which is ultimately determined by the physical environment. Environmental variables such as the amount of available oxygen, water temperature and rate of sedimentation control the ultimate fate of a community of organisms. Building up a clear picture of fossil communities in the geological record is difficult since few soft-bodied organisms are preserved and because accumulation of fossils usually takes place over a considerable time period, during which a community may change. Despite these limitations it is possible to reconstruct ancient fossil communities in some locations and as a result one can deduce the environmental parameters which limited them. Consequently the distribution of fossil communities should provide important information with which to reconstruct ancient palaeogeographies. The work of

Ziegler (1965) provides a very clear illustration of the application of this idea in the Lower Palaeozoic rocks of Wales.

Ziegler's (1965) work is based on the detailed observation of fossil assemblages (biofacies) present in the rocks of the Lower Silurian (Llandovery) in Wales and the Welsh Borders. The area was located on the southern margin of a contracting ocean, the Iapetus Ocean, during the Silurian. The rocks consist of turbidite and dark mud facies in the north-west and a sandier facies to the south-east. Ziegler found that there were five recurring assemblages of bottom-dwelling organisms. The contents of these assemblages were dominated by brachiopods. The detailed nature of each assemblage was described in full in a later paper (Ziegler *et al.* 1968). Brachiopods are marine, shelled organisms with a delicate feeding mechanism and are not widely distributed today. In the Early Palaeozoic, however, they were amongst the commonest type of sea floor organism. Ziegler recognised each of his five assemblages or communities on the basis of common species, as well as on the relative abundance of individual species within the community (Figure 7.8). In the 1920s the Danish ecologist Petersen had discovered that it was possible to map the relative distribution of marine communities in the shallow seas of today. He determined that they form distinct patterns reflecting the distribution of marine environments. On this basis Ziegler (1965) argued that the recurrent set of fossil communities recognised in the rocks of the Llandovery in Wales reflected the distribution of marine environments along the southern margin of the Iapetus Ocean.

Ziegler mapped the geographical distribution of his five fossil communities in the Llandovery outcrops of Wales and the Welsh Borders. From this it became apparent that the five communities fell into a number of arcuate, concentric zones at any point in time and that their relative position remained constant. For example, if one was to move progressively westwards across the area the following communities would be encountered: *Lingula, Eocelia, Pentamerus, Stricklandia* and *Clorinda*, before one moved into the basinal muds in which there were few bottom-living organisms. The *Lingula* community borders the upstanding Precambrian massif of the Malvern Hills in central England and Ziegler (1965) argued that the succession of fossil communities represented a progression offshore: the communities changing with increasing water depth and decreasing water turbulence. In this way the arcuate pattern of communities could be interpreted in terms of an ancient shoreline (Figure 7.8).

Ziegler (1965) developed this concept further by mapping the distribution of these five fossil assemblages through time as well as space. In this way he was able to track the movement of the coastline inshore during the Llandovery. The vertical changes in the type of communities present with the rock sequence reflect the progressive migration of deeper water coastal zones onshore (Figure 7.8).

2. **Comment**

This case study illustrates the application of uniformitarian principles to the interpretation of biofacies. Given that fossil communities can be recognised, the interpretation of the mode of life of the animals within a biofacies enables a reconstruction of the environment in which they lived and in some cases even of the palaeogeography.

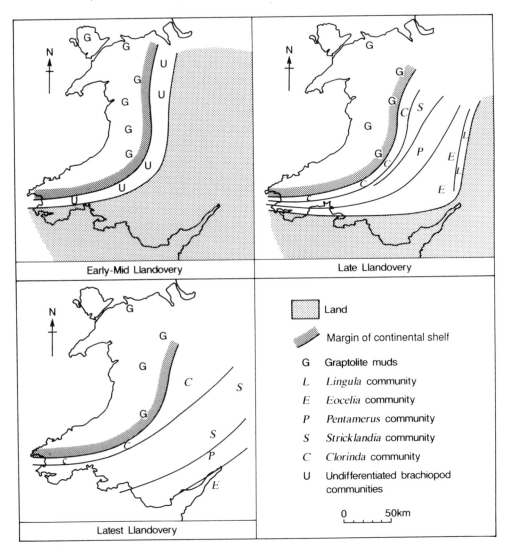

Figure 7.8 *Early Palaeozoic shorelines in Wales deduced from biofacies. The shorelines plotted on the maps were constructed from the distribution of fossil assemblages. The maps depict a progressive marine transgression during the Llandovery. [Modified from: Ziegler (1965) Nature* **207**, *Figs 1–3, pp. 270–271]*

3. References

Ziegler, A. M. 1965. Silurian marine communities and their environmental significance. *Nature* **207**, 270–272.

Ziegler, A. M., Cocks, L. R. M. & Bambach, R. K. 1968. The composition and structure of Lower Silurian marine communities. *Lethaia* **1**, 1–27.

7.3 THE APPLICATION OF STRATIGRAPHY TO PLATE TECTONICS

Plate tectonics has provided an important model with which to interpret the stratigraphical sequence. However, the application of simple stratigraphical tools has also played an important role in refining the plate tectonic model. The case study chosen in this section tells the story of the recognition and development of the concept of terranes. As outlined in Box 6.2, terranes are produced by the accretion of different 'slivers' of crust onto a continental margin during a process known as collage tectonics. The initial recognition of terranes in the North American Cordilleran mountains was based upon simple stratigraphical analysis of the Cordillera, which displays widely differing stratigraphies and fossil faunas in close juxtaposition, separated by faults.

7.3.1 Allochthonous Terranes
(Irwin 1960, 1972; Monger & Ross 1971; Coney *et al.* 1980; Tipper 1984)

1. **Introduction**
   Tectonic belts are linear belts of deformed crust. Before the advent of plate tectonic theory these belts were interpreted as geosyclines or regional downwarps in which sediment accumulated and was deformed. With the advent of plate tectonic theory these linear belts were reinterpreted in terms of continental collison and the Wilson Cycle (Figure 6.6). This model of mountain building has developed further in recent years with the recognition of **allochthonous terranes** within many tectonic belts. Allochthonous terranes are fault-bounded tracts of land each with a distinct stratigraphical sequence and fauna. Today many mountain belts are interpreted in terms of terranes and are believed to have formed by the accretion of a variety of different 'slivers' of crust (terranes) which are accreted either between colliding continents or along the edge of a continent to produce mountain belts (Figure 6.11). The initial recognition of allochthonous terranes owes much to the field observation of different stratigraphies and therefore geological histories as well as the recognition of unusual or foreign faunas. The concept of allochthonous terranes was developed in the North American Cordilleran Mountains, where in excess of 50 terranes have been recognised. This work, and the development of the terrane concept, has had a profound effect on the interpretation of tectonic belts.

2. **Recognition of disjunct faunas and the discovery of terranes**
   Monger and Ross are palaeontologists who in the early 1970s were studying the distribution of the fossil remains of large single-celled organisms (foraminifera) called fusulinids. Fusulinids lived on the sea floor and were common in the Late Palaeozoic, becoming extinct during the Late Permian. In studying the Permian rocks of British Columbia Monger and Ross recognised that in the area around Cache Creek the fusulinid fauna was of European aspect, with the species present common only in Permian marine rocks of the European Alpine belt. This contrasted with areas to the west and east of Cache Creek which contained faunas of North American affinity. The implication of this was that the Cache Creek region was

actually displaced and allochthonous to the region. This work revived interest in observations made much earlier by Irwin (1960).

Irwin (1960) had recognised that the Klamath Mountains in northern California could be divided into four distinct belts on the basis that each belt had a distinct fauna and that the nature of the stratigraphy in each belt was different. In his later work Irwin showed that these belts were fault bounded and that each sequence was therefore a separate **tectono-stratigraphical entity** brought together by tectonic activity, but not having been originally in close juxtaposition (Irwin 1972). This led to Irwin's (1972) definition of a terrane as 'an association of geological features, such as stratigraphical sequences, palaeontology, intrusive rocks, mineral deposits and tectonic history which differ from those of an adjacent terrane'.

3. **Wider implications**

Coney *et al.* (1980) reviewed the body of evidence about allochthonous terranes that had been growing from the initial palaeontological and stratigraphical data. They recognised in excess of 50 terranes accreted onto the ancient rocks of the North American craton (Figure 7.9). They recognised terranes on the basis of their internal homogeneity and the continuity of stratigraphy, tectonic style and geological history within their boundaries. The terrane boundaries separate distinct stratigraphies within the terranes and juxtapose different faunas. In this way the recognition of individual terranes is based solely on the stratigraphical character of the terrane. A terrane is first recognised on the basis of a stratigraphical history sufficiently different from that of an adjacent terrane and distinct enough not to be a result of a facies change. Many of the terranes recognised by Coney *et al.* (1980) have stratigraphical histories indicative of an oceanic setting, with oceanic sediments and basaltic lavas, while others seem to represent fragments of volcanic arcs or small continental fragments. In turn these are distinguished from the Palaeozoic and pre-Palaeozoic stable craton of the North American interior. The Late Mesozoic history of these terranes shows that they were accreted during this period onto the Cordilleran margin. The date at which each terrane docked with the North American craton can be determined by an analysis of cross-cutting relationships and the dating of igneous bodies of rock associated with the docking of a particular terrane.

Coney *et al.* (1980) postulated that the origin of the terranes was connected with the closure of the Pacific Ocean, which was dominant in the Late Palaeozoic to Early Mesozoic when most continental crust was part of the large landmass of Pangaea. Island arcs and fragments of oceanic crust were swept from the west against the cratonic margin and moved laterally along it to produce the mosaic of terranes that we see today.

4. **The importance of biogeography to terrane analysis**

Tipper (1984) demonstrated that the distribution of fossil organisms is a strong tool in the interpretation of the stratigraphical record of the Cordilleran terranes. He found that the ammonite faunas of the Canadian Cordillera could be used to show the relative lateral displacement of terranes both before and after they docked with the North American continent (Figure 7.10).

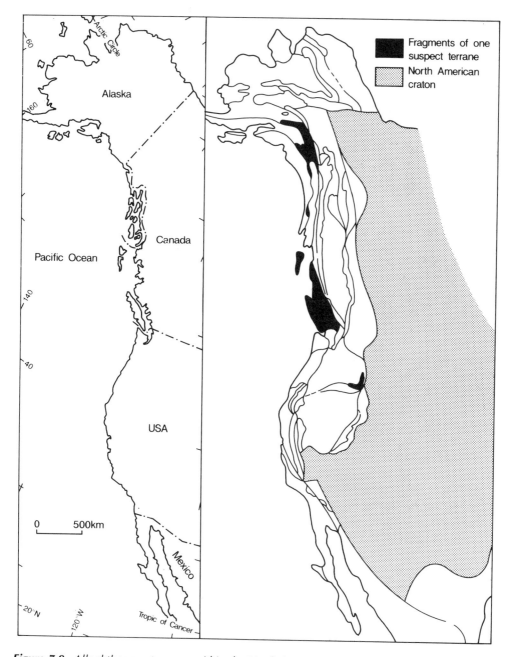

Figure 7.9 *Allochthonous terranes within the North American Cordillera. The fragments of one suspect terrane are picked out in black. [Modified from: Coney et al. (1980) Nature **288**, Fig. 1, p. 330]*

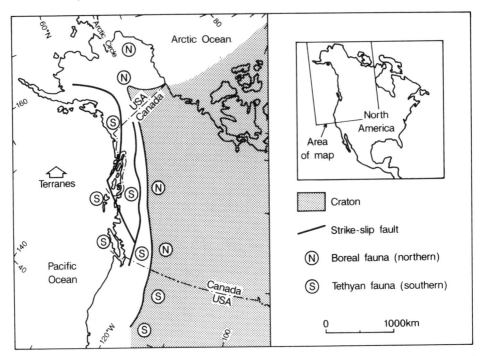

Figure 7.10 *The use of faunal realms to deduce the lateral movement of terranes within the Canadian Cordillera. The boundary between the Boreal and Tethyan faunal realms has been displaced northwards in the area of accreted terranes. [Modified from: Tipper (1984) In: Westermann (Ed.)* Jurassic–Cretaceous Biochronology and Palaeogeography of North America, The *Geological Association of Canada Special Paper 27, Fig. 3, p. 115]*

It is well established that, in common with many other Jurassic marine organisms, ammonite species are organised into distinct latitudinal belts which may reflect the presence of the main climatic belts on the Earth's surface. In the later part of the Early Jurassic the ammonites became segregated into latitudinal faunas, usually referred to as the Boreal (northern) and Tethyan (southern) realms. These realms can be identified in the rocks deposited on the flooded craton at this time as well as in the Canadian terranes of Wrangellia, Stikinia and Quesnellia. It is clear that these three terranes had differing stratigraphical histories until the Middle Jurassic, and that in the Late Jurassic all three terranes had moved northwards along faults which bounded each terrane. This was apparent because the latitudinal position of the faunal realms in the three terranes has been offset and is at odds with the distribution of the faunal realms on the stable craton. Docking and the lateral displacement of these three terranes were completed by the Middle Jurassic, since they are stratigraphically and faunally similar thereafter.

5. **Comment**
   The discovery of terranes has had a profound effect on how we view tectonic belts. Terranes are identified on stratigraphical and faunal evidence within many of the

Earth's tectonic belts; and their recognition increased our understanding of tectonic processes. As we will show in Chapter 11, the British Isles is viewed by many people as a series of terranes which were brought together during the Caledonian Orogeny in the Palaeozoic. The case history illustrates that stratigraphical analysis of lithostratigraphy, biostratigraphy and facies are powerful tools in plate tectonic modelling.

6. **References**

Coney, P. J., Jones, D. I. & Monger, J. W. H. 1980. Cordilleran suspect terranes. *Nature* **288**, 329–333.

Irwin, W. P. 1960. Geological reconnaissance of the northern Coast Ranges and Klamath Mountains, California. *Californian Division of Mines Bulletin*, **179**.

Irwin, W. P. 1972. Terranes of the western Paleozoic and Triassic belt in the southern Klamath Mountains, California. *US Geological Survey Professional Paper* **800-C**, 103–111.

Monger, J. W. H. & Ross, C. A. 1971. Distribution of fusulinaceans in the western Canadian Cordillera. *Canadian Journal of Earth Sciences* **8**, 259–278.

Tipper, H. W. 1984. The allochthonous Jurassic–Lower Cretaceous terranes of the Canadian Cordillera and their relation to correlative strata of the North American craton. In: Westermann, G. E. G. (Ed.) *Jurassic–Cretaceous Biochronology and Paleogeography of North America*. Geological Association of Canada Special Paper No. 27, 113–120.

7.4 SUMMARY OF KEY POINTS

In this chapter we have brought together several case histories to illustrate the basis of stratigraphical procedure, and the interpretation of the rock succession in terms of a sequence of events in Earth history. These case studies all illustrate that stratigraphy involves three basic elements: **observation, description** and **interpretation**. In establishing the rock succession the stratigrapher observes and describes the rock units present. When dating the sequence, using either relative or absolute techniques, the stratigrapher establishes a chronostratigraphical framework for the units he or she has described. The description of units is vital for their interpretation. Interpretation of rock units in terms of ancient environments, ecologies and geographies is underpinned by the principle of uniformitarianism. Most geologists accept that plate tectonics is the unifying concept in geology but, as the last case study has shown, the plate tectonic model is frequently refined. In this case allochthonous terranes were recognised through the simple observation, description and interpretation of stratigraphical sequences. The implications of these simple observations have caused the plate tectonic model to be refined which has in turn refined our interpretation of tectonic belts.

8
Summary of Part I:
The Stratigraphical Tool Kit

In the first part of this book we have examined the stratigraphical tools necessary to fulfil the main aim of stratigraphy, which is the interpretation of sequences of rock as a series of events in the evolution of the Earth. In order to do this there are three tasks which must be undertaken: (1) to establish the sequence of rock units present; (2) to work out the chronological order of those rock units; and (3) to interpret the environment in which each unit was deposited or formed, so that a record of the changing environment can be established. Since the 1960s plate tectonics has provided the broad theoretical framework with which to view the global development of the stratigraphical record and therefore a fourth task is to determine the global perspective with a stratigraphical sequence in the light of plate tectonic theory. Part I has introduced you to the basic tools which enable you to carry out these tasks and fulfil the aim of stratigraphy. A recap of these tools is provided below.

Actualism—the modern interpretation of uniformitarianism—recognises that the laws of nature have remained unchanged throughout most of geological time, although the rates at which such processes operate may have varied. The concept of actualism underpins our understanding and interpretation of the geological record. It is the most fundamental principle in stratigraphy and the most important element of the stratigraphical tool kit.

The special component of geology which sets it apart from almost all other sciences is that of time. It is important to determine the **relative chronology** or timing of the sequence of rock units. In bedded sequences, **superposition** determines that the lowest layer in an undeformed sequence is the oldest. In non-bedded sequences, the cross-cutting of rock masses by faults or igneous bodies, for example, helps to determine the relative chronology. **Absolute geological time** may be determined by the radioactive decay of unstable isotopes, but in practice the relative order of events is sufficient for most geologists. The relative order may be determined by the position of evolving fossils or through the recognition of **instantaneous events**. The **Chronostratigraphical Scale** is important as it provides a global standard for correlation of geographically or environmentally different sequences.

The rock record is interpreted for the most part through the application of actualism and the recognition that the processes which have operated can be determined from the characteristics of the product. In this way, the sum total of the characteristics of a rock body, called its **facies**, can be used to interpret the environment prevailing during its formation or deposition. The recognition of facies logically leads to the construction of a **palaeogeography**, each facies corresponding to an environment which may be plotted on a map to represent the geography of an area at any one time. Climate has a profound effect on sedimentation: the prevailing palaeoclimatic conditions at the time of deposition can therefore be inferred from the characteristics of many sedimentary rocks. Fossils too provide an important insight into the development of an environment at any one time and **palaeoecologies** can be reconstructed.

Part I of this book is therefore about the ways and means of obtaining detailed information from the stratigraphical record. In most cases this means the collection of local information from local rock sections. The work of countless geologists has contributed in this way to the recognition of global pattern in the rock record in both space and time. Part II illustrates the record of global change determined from the tools outlined in the previous pages and shows how these changes have interacted to produce the Earth's stratigraphical record.

Part II
THE PATTERN OF EARTH HISTORY

9
Introduction to Part II

As we have shown in Part I, stratigraphy is the interpretation of rock successions in terms of sequences of events in Earth history. Each stratigraphical unit contains a number of clues to the nature of the environment in which it was formed. By piecing together these clues it is possible to determine the nature of the changing environments of the Earth through time. Stratigraphy therefore provides the temporal framework within which we can compare ancient environments and study the way in which the Earth has changed.

The traditional approach to stratigraphy is to build up a pattern of Earth history brick-by-brick, layer-by-layer. Only when all the layers are in place can the global pattern and large-scale controls on the development of the stratigraphical record be appreciated. In this part of the book we shall attempt to provide an alternative approach in which we shall build up a framework of the global controls on the development of the stratigraphical record and their variation through geological time. We shall do this before illustrating how this framework can be used to view the geological development of one small part of the Earth's crust, in this case the British Isles.

What are the global or large-scale controls on the stratigraphical record? Today, the Earth's physical environment is the product of the interaction of four variables: **plate tectonics**, **climate**, **sea level** and the **biosphere**. If one applies the principle of uniformitarianism then the ancient environments deduced from each stratigraphical unit can also be interpreted in terms of these four variables. Together they give a record of Earth history. These four variables control the nature of the depositional environment and together have created the stratigraphical record. Within these variables there is a distinct hierarchy of importance (Figure 9.1). Plate tectonics is at the head of this hierarchy and is fundamental because it controls the formation and development of depositional basins within which the sediments, and therefore the stratigraphical record, can accumulate. Plate tectonics also determines the production and distribution of igneous and metamorphic rocks and the nature of the tectonic processes which

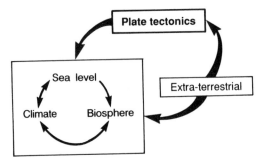

Figure 9.1 *Flow diagram showing the interaction of the key variables which determine the nature of the physical environment of the Earth*

deform the stratigraphical record. In contrast, climate, sea level and the biosphere, although having a profound influence on the nature of the sedimentary environment, only control the nature of the sedimentary facies preserved in the depositional basins created by plate tectonics. These three variables are therefore strongly dependent on plate tectonics and are also closely interrelated with one another. In this way, the interaction of the four variables is responsible for the production of the Earth's stratigraphical record. It is, however, important to note that the Earth may also have experienced extra-terrestrial influences, particularly the impact of meteorites and other similar bodies. Extra-terrestrial impact has frequently been invoked to explain changes within the biosphere and climate and also in plate tectonic activity, although for the most part these hypotheses remain speculative.

In Part II, Chapter 10 explores the way in which plate tectonics, sea level, climate and the biosphere have varied globally throughout Earth history, each with their own independent but interrelated patterns of temporal change. This exploration is only possible because of the detailed investigations of generations of 'stratigraphical detectives' who have pieced together the sequence of events and interpreted their significance. In Chapter 11 the record of global change is examined with reference to the stratigraphical record of one part of the Earth's crust, the British Isles.

10
The Stratigraphical Record and Global Rhythm

The stratigraphical record can be interpreted in terms of four variables: plate tectonics, climate, sea level and the biosphere. These four variables combine to determine the nature of the rocks deposited at any point on the surface of the Earth. Throughout the Earth's history these four variables have varied systematically through time, each with a different rhythm. In this chapter the rhythm of change through time of each of these four variables is explored.

10.1 PLATE TECTONICS

Plate tectonics controls the evolution, shape and closure of sedimentary basins and is therefore probably the most important control on the development of the stratigraphical record. Deposition may occur widely but the preservation of sediments is largely controlled by the availability of basins in which they can accumulate. Without such basins there would be little sediment preserved and therefore a very poor stratigraphical record. Plate tectonics is the means by which the stratigraphical record is written, or preserved, while by contrast variables such as sea level and climate simply determine the character of that record, or, to continue the analogy, they only determine the colour of the ink with which it is written. It is, however, important to remember that not all volcanic or tectonic activity can be accounted for and be predicted by the plate tectonic model and that occasionally intra-plate volcanic and tectonic events may occur, associated with global 'hot spots'.

10.1.1 Plate Tectonics Through Time

Since the introduction of the plate tectonic model in the 1960s it has been established that plate tectonic processes have been operative throughout the Phanerozoic and probably in some form during the later part of the Precambrian. Plate tectonics has provided an important key to understanding the Earth's stratigraphical record, but how far back can the model be applied?

The Earth's interior has been cooling since its formation some 4600 million years ago. During the Archaean the mantle was much hotter and consequently the Earth's crust would have been warmer and therefore more thermally buoyant than at present. It has been suggested that this increased buoyancy may have prevented the subduction of ocean crust in the way predicted by the conventional plate tectonic model. As the mantle cooled subduction became possible and conventional plate tectonics took over. The question of whether the Archaean crust could subduct is open to debate and some would argue that a slightly different tectonic regime may have prevailed at this time. The plate tectonic model can, however, be used with some confidence to explain the stratigraphical record during the later part of the Precambrian, or the Proterozoic, and throughout the Phanerozoic.

During the Proterozoic (Late Precambrian) and Phanerozoic intervals the position of the Earth's continents at any given time is difficult to establish with absolute precision, but a picture of continental distribution can be obtained by a study of palaeomagnetic data. These data can be used to obtain an estimate of the former position of continents and therefore provide a tool with which the changing pattern of the Earth's continents through time can be established, at least for the Phanerozoic.

The changing pattern of continental blocks during the Phanerozoic is illustrated in Figure 10.1. This pattern of changing continents is perhaps best viewed as a cycle in which the continents were initially widely dispersed at low latitudes in the Early Palaeozoic, before they coalesced in the Late Palaeozoic to form a single supercontinent (**Pangaea**). This supercontinent was again fragmented in the Mesozoic to give the dispersed pattern of continents present today (Figure 10.1). It has been suggested that a similar **supercontinental cycle** of continental dispersal–aggradation–dispersal can be identified in the Proterozoic. Continental aggradation is associated with mountain building or **orogenesis**.

Recent work has suggested that such supercontinental cycles may result from the contrast in the thermal properties of continental and ocean crust. Continental crust is only about one-half as efficient at conducting heat as oceanic crust, so that if a supercontinent such as Pangaea covers a significant part of the Earth's surface, then heat will build up in the mantle below it. Given time the build up of heat beneath the supercontinent may lead to uplift, rifting and continental dispersal. As ocean basins formed during continental dispersal become mature, subduction zones will develop. This will lead to a transition from continental dispersal to continental aggradation and the development of another supercontinent. Each time a new supercontinent is assembled ocean basins close and mountains are built.

In summary, the evidence suggests that the world's continents have experienced a period of assembly and dispersal during the Phanerozoic which has had a profound effect on the geological record of this period. A similar cycle may also have occurred in the Proterozoic.

10.2 CLIMATE

The Earth's climate has varied throughout geological time. This is clear from the evidence provided by the stratigraphical record which we discussed in Section 5.3. However, the mechanisms by which climate has changed are poorly understood and numerous terrestrial as well as extra-terrestrial factors have been proposed to explain the Earth's climatic record. These explanations include variation in topography, sea level, continental distribution, atmospheric carbon dioxide, solar radiation and the position of the Earth's solar system within the galaxy.

10.2.1 Climate Through Geological Time

Using the palaeoclimatic tools outlined in Section 5.3 it is possible to obtain an impression of the climate which existed during a particular geological time interval. This is done by plotting the distribution of palaeoclimatically sensitive lithologies or fossils of a similar age on a palaeogeographical map of the period in question. When this information is combined with other sources of evidence, such as chemical isotope data, a reconstruction of the palaeoclimate can be obtained. A broad picture of the Earth's palaeoclimate over the last 500 million years has been established in this way, but our knowledge of Precambrian climates is poor, since palaeoclimatic reconstruction becomes difficult with increasing age.

During the Late Proterozoic and Phanerozoic the Earth's climate appears to have oscillated between two stable states: one of global warming, the 'Green-house' state, and one of global cooling, the 'Ice-house' state. During the periods of global refrigeration, or 'Ice Ages', ice sheets dominated the high-latitude areas of the globe and left a clear record of their presence in the form of deposits and landforms. During 'Green-house' periods the Earth was usually ice free and there was generally a period of global warmth. The number of oscillations between these two states detectable in the stratigraphical record depends largely on the scale or detail with which the record is viewed.

On the broadest scale it is possible to recognise seven 'Ice-house'–'Green-house' oscillations in the Phanerozoic, and we review these, and some of the evidence to support them, below. In addition, despite the fact that current knowledge of Precambrian palaeoclimate is poor, there is good evidence to suggest that the Earth was in an 'Ice-house' state during the Proterozoic. This glaciation appears to have been one of the most extensive and long-lived periods of glaciation in the Earth's history and forms an important horizon or event from which to begin a brief review of the Earth's palaeoclimatic history.

1. **Ice-house world: Late Proterozoic (800–570 million years ago).** This period of glaciation is one of the most extensive within the geological record and may have lasted for up to 230 million years. It is particularly puzzling because evidence of glaciation is not only found in rocks which formed at high latitudes, but also in those which formed at low latitudes of less than 45°. This period appears, therefore, to have been a particularly intense episode of global cooling.

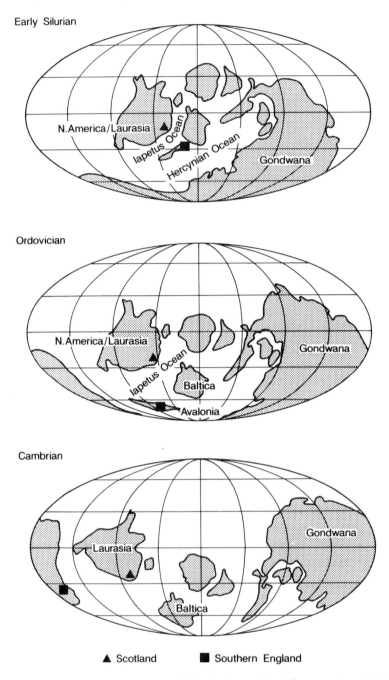

Figure 10.1 *The distribution of continental blocks through the Phanerozoic. It is important to note that areas of these continental blocks were flooded at various times. [Based on information in: Scotese (1991)* Palaeogeography, Palaeoclimatology, Palaeoecology *87, 493–501; Scotese & McKerrow (1990)* Memoir of the Geological Society of London *12, 1–21]*

Early Permian

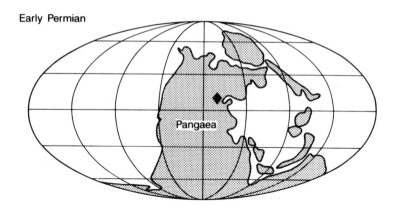

Late Carboniferous

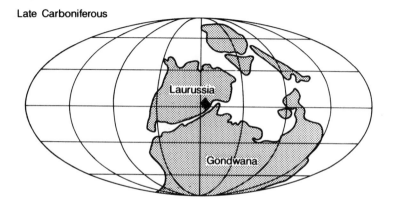

Early Devonian

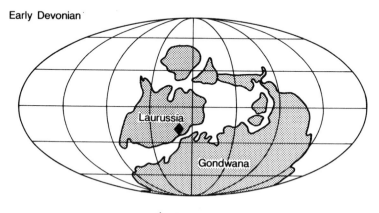

◆   British Isles

Mid-Cretaceous

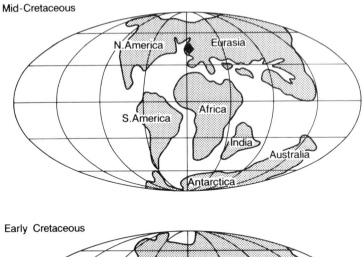

Early Cretaceous

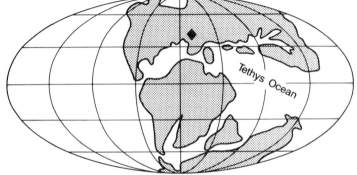

Jurassic

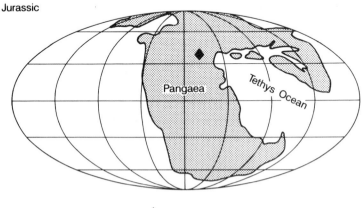

◆ British Isles

2. **Green-house world: Early Cambrian to Late Ordovician (570–458 million years ago).** During the Early Cambrian the Earth began to warm. The process of global warming appears to have been slow and reached a peak about 468 million years ago. Average low-latitude temperatures at this time have been judged to be about 40°C, with the climate a little wetter than at present.

3. **Ice-house world: Late Ordovician to Early Silurian (458–428 million years ago).** The first glaciers appear to have formed just after 458 million years ago and developed into a full ice sheet glaciation around 10 million years later in northern Africa, which was then situated close to the South Pole. Local, smaller ice centres developed in other parts of Africa and South America as the continent of Gondwana moved over the South Pole. The average low-latitude temperature has been judged to be about 22°C, which is not too dissimilar to the present-day average of 24°C. Other parts of the Earth appear to have experienced arid conditions during this period.

4. **Green-house world: Early Silurian to Early Carboniferous (428–333 million years ago).** This 'Green-house' period started with slow gradual warming which persisted until about 367 million years ago. This period of warming is associated with an irregular increase in aridity which is illustrated by an abundance of evaporites and an absence of the humid indicators, coal and bauxite.

5. **Ice-house world: Early Carboniferous to Late Permian (333–258 million years ago).** This Late Palaeozoic cool mode was associated with the formation of high-latitude glaciers on many parts of the continent of Gondwana, which was located in the Southern Hemisphere. Early glaciers grew into ice sheets and persisted for some 65 million years. The ice sheets reached their greatest extent between 315 and 296 million years ago when they reached a latitude of 35°S. Glaciation was episodic and it has been suggested that each episode is recorded in the sea level fluctuation that controlled the deposition of the coal-rich sediments of the Northern Hemisphere which were laid down at this time. The cyclic expansion and contraction of ice during the Carboniferous is similar to that recorded during the recent Quaternary 'Ice Age' and is probably due to Milankovitch radiation variations (Box 10.1). There is evidence to suggest that beyond the glaciated areas the climate was more humid than at present.

6. **Green-house world: Late Permian to Early Cenozoic (258–55 million years ago).** This has been generally regarded as an extensive period of global warmth when conditions were considerably more equable than at present. However, recent work has suggested that within this long period of warmth a cool mode can be identified, although not of the severity of previous episodes. One can therefore subdivide this long period into three shorter ones. First, between the Late Permian and Middle Jurassic (258–187 million years ago) there was distinct warm mode. The evidence for this is poor, but it has been suggested that the period was one of considerable warmth. The influence of the large continental landmass of Pangaea may have been important at this time, with extensive areas of this huge landmass in which rainfall was very low and temperatures very high.

Second, between the Middle Jurassic and middle part of the Cretaceous (187–105 million years ago) there was a period of cooler, and more seasonal, conditions. The

BOX 10.1: MILANKOVITCH CYCLES

In 1941 the astronomer Milutin Milankovitch calculated the magnitude and frequency of changes in the solar radiation received by the Earth as a consequence of cyclic changes in its orbit. He identified three orbital processes which would control these changes. (1) **Axial tilt**: the Earth's axis of rotation is inclined at an angle of about 23.5°. This angle is not constant and varies between 24.5 and 21.5° every 41 000 years. When the angle of tilt is greatest there is the greatest difference in seasonal heating at any latitude. (2) **Eccentricity of the orbit**: the Earth's orbit changes from an elliptical orbit to a circular one every 100 000 years. (3) **Precession of the equinoxes**: the Earth wobbles slightly on its axis due to the gravitational pull of the Sun and Moon; consequently the time of the year at which the Earth is nearest to the sun changes through time in a cycle which takes about 23 000 years. Eleven thousand years ago the Earth's nearest point to the sun occurred when the Northern Hemisphere was tilted towards the sun (Northern Hemisphere summer), rather than during the Northern Hemisphere's winter as is the case today. These three orbital cycles cause subtle variations in the amount of solar radiation received at different latitudes through time. It is these variations which provide the pulse beat of glacial expansion and contraction during 'Ice Ages'.

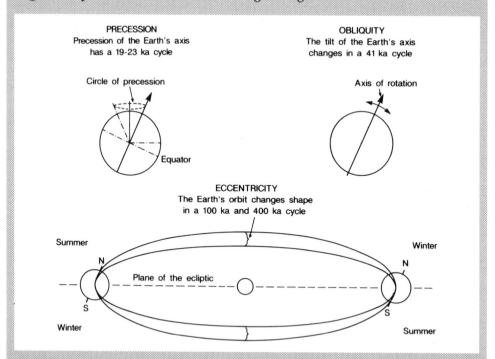

Source: Imbrie, J. & Imbrie, K. P. 1979. *Ice Ages: Solving the Mystery*. Harvard University Press, Cambridge, Massachusetts. [Diagram reproduced by permission from: Boulton (1993) In: Duff (Ed.) *Holmes' Principles of Physical Geology*, Chapman & Hall, Fig. 21.9, p. 447]

BOX 10.2: DROPSTONES IN THE CRETACEOUS

The Mesozoic is traditionally considered to have been one of the longest 'Greenhouse' periods within the Phanerozoic, with global warmth and ice-free poles. Recently, the work of Frakes and Francis (1988) has challenged this model. The Lower Cretaceous Bulldog Shales of the intracratonic Eromanga Basin of central Australia are laminated marine mudrocks which contain large rock clasts of up to three metres diameter. The question raised by these clasts is this. Laminated mudstones are commonly produced in still waters, while transport of such outsized clasts requires very swift currents. The mudrocks do not contain any evidence of strong current activity, so what is the agent of transport? Frakes and Francis suggest that the clasts must have been dropped into the mud from a raft, and conclude that the most likely rafting agent is that of seasonal sea ice, which picks up clasts in the winter and releases them in the summer. This has implications for the supposedly warm climate of the Cretaceous, which according to these conclusions must have had at least seasonal ice in central Australia during the Cretaceous (palaeolatitude 65° and 78°S). This nicely illustrates the power of small facies interpretations in determining global climatic patterns. There is, however, one word of caution. In his voyage on HMS *Beagle*, Darwin observed that 'the inhabitants of the Radack archipelago, a group of lagoon-like islands in the midst of the Pacific, obtained stones for sharpening their instruments by searching the roots of trees which are cast upon the beach.' (Darwin 1879, p. 461)

Sources: Frakes, L. A. & Francis, J. E. 1988. A guide to Phanerozoic cold polar climates from high-latitude ice-rafting in the Cretaceous. *Nature* **333**, 547–549. Darwin, C. 1879. *Naturalists' Voyage Round the World*. (Second Edition) John Murray, London.

recent discovery of large ice-rafted dropstones in strata of this age may indicate that seasonal ice may have existed during this period at high latitudes (Box 10.2), but as yet no direct evidence of glaciation during this period has been recorded.

Finally, following this cooler period there was a return to a warm humid climate between the middle part of the Cretaceous and the Early Cenozoic (105–55 million years ago). This period was one of the warmest of the Phanerozoic with the average global temperature probably about 6°C higher than that of today. Temperatures were high enough for the poles to be completely ice free and home to both forests and large reptiles (in the Cretaceous). The mid-Cretaceous was also a time of globally high sea levels and extensive areas of shallow shelf seas, ideal for the development of moderate climates, due to the increased evaporation and precipitation. During the Late Cretaceous sea levels were generally lower and the climate appears to have been more seasonal and continental.

7. **Ice-house world: Early Cenozoic to Present (55 million years ago to the present day).** The first stages of the Cenozoic cool mode started with global cooling in the Early Cenozoic (Eocene) some 55 million years ago. From this time the Earth's

climate gradually cooled from the warm conditions of the Mesozoic to the cool glacial climates of today. The Earth cooled throughout the Cenozoic and polar ice caps developed in Antarctica and Greenland. This period of cooling was followed, two million years ago, by the growth of mid-latitude ice sheets in both Scandinavia and North America. Most of the British Isles was glaciated and much of the current landscape was formed during this period. These mid-latitude ice sheets have accumulated and decayed in a cyclic fashion throughout the last two million years. Periods of ice expansion are known as **glacials,** while periods in which the mid-latitudes were ice free are known as **interglacials**. The Earth is currently within an interglacial and a glacial interval, along with mid-latitude ice sheets, is likely to return within the next few thousand years. This cyclic expansion and contraction of mid-latitude ice sheets during the Quaternary is due to systematic variations in solar radiation caused by changes in the Earth's orbit, known as Milankovitch radiation variations (Box 10.1). A detailed record of the growth of the polar ice caps during the Cenozoic and of the cyclic expansion and contraction of the mid-latitude ice sheets during the Quaternary is recorded within the oxygen isotope record (Box 5.3).

One can identify several similarities between each 'Ice-house' and 'Green-house' state within the stratigraphical record. 'Ice-house' states appear to start with a long period of cooling which culminates in the growth of high-latitude ice sheets. Such periods are normally terminated by a sudden episode of climatic warming. They are invariably associated with the presence of extensive continental landmasses at high latitudes. In fact the migration of the centre of glaciation between the Ordovician and Carboniferous glaciations can be shown to reflect the movement of the Gondwana supercontinent (Figure 10.2). Within 'Ice-house' periods the occurrence of mid-latitude ice sheets is uncommon. However, they are particularly a feature of the Quaternary 'Ice Age', and possibly also occurred in both the Carboniferous and Late Proterozoic 'Ice Ages'. The variation in intensity of 'Ice-house' periods is hard to explain, but the presence of mid-latitude ice sheets in the Quaternary 'Ice Age' is probably a function of Cenozoic mountain building (Box 10.3). In contrast, 'Green-house' conditions appear to start rapidly and are normally terminated in a more gradual fashion via a period of cooling which takes place at different rates in different geographical locations. 'Green-house' conditions are loosely correlated with periods of geological time when: (1) the atmosphere possessed higher concentrations of green-house gases, such as carbon dioxide; (2) sea level was high and therefore large areas of the continents were flooded; and (3) the Earth's ocean basins were either very large or well connected.

In summary, the Earth's climate appears to have oscillated between periods of global cooling ('Ice-house') and global warming ('Green-house'). It is important, however, to note that superimposed on this large-scale oscillation are smaller-scale temperature fluctuations within a given warm or cold phase. This scale of variation is well illustrated by the growth and decay of mid-latitude ice sheets during the Quaternary. It is important to note that the existence of an 'Ice-house' period does not necessarily imply that all of the Earth was glaciated, but simply that the climate was cooler and that the equatorial belt may have been more restricted.

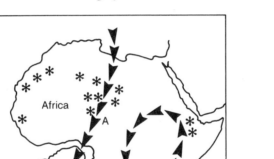
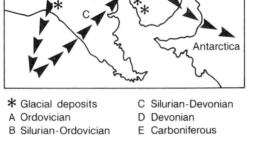

* Glacial deposits   C Silurian-Devonian
A Ordovician   D Devonian
B Silurian-Ordovician   E Carboniferous

Figure 10.2 *The movement of the centre of glaciation (black arrows) over the supercontinent of Gondwana, from the Ordovician to the Carboniferous. This probably reflects the movement of the supercontinent over the South Pole during this period. [Modified from: Boulton (1993). In: Duff (Ed.)* Holmes' Principles of Physical Geology, *Chapman & Hall, Fig. 21.29, p. 466]*

The change from one climatic state to another must have been associated with enormous changes in the climate system, while the relative homogeneity of each mode, persisting over tens of millions of years, indicates a large degree of inertia to change within both climatic systems. What causes this large-scale oscillation in climate? One can identify four broad types of climatic forcing which may be responsible: (1) extra-terrestrial causes; (2) variation in the amount and distribution of solar radiation received at the top of the atmosphere; (3) variation in the composition of the atmosphere; and (4) the nature of the Earth's surface—the distribution of land and sea.

1. **Extra-terrestrial causes.** The impact of meteorites or 'bolides' has been put forward as a possible explanation for the change between 'Ice-house' and 'Green-house' states. The impact of a large meteorite could potentially cause global cooling through the production and injection of masses of dust into the upper atmosphere, blocking off the sun's radiation. The impact of a large extra-terrestrial body in an ocean might also promote higher temperatures through the vaporisation of large

BOX 10.3: WHY DO MID-LATITUDE ICE SHEETS FORM IN SOME 'ICE AGES' BUT NOT IN OTHERS?

Certain 'Ice Ages' in the geological record appear to have been more intense than others. For example, the Cenozoic 'Ice-house', and possibly the Carboniferous and Late Proterozoic 'Ice-houses', were associated with mid-latitude ice sheets, while other episodes of 'Ice-house' conditions experienced only polar glaciation. It has been suggested that the intensity of an 'Ice Age' is determined by the pattern of atmospheric circulation. The Cenozoic 'Ice-house' is believed to have been intensified by the modification of atmospheric circulation through mountain building during the Cenozoic. Within the Northern Hemisphere there are two regions of extensive highland: (1) the Tibetan Plateau and the Himalayan mountains of Asia and (2) the Western Cordilleras of North America. Both these areas are broad, plateau-like bulges on which narrower mountain ranges are superimposed and they only reached their present elevation within the last million years. The large-scale circulation of the atmosphere is strongly influenced by these areas of uplift (see diagram opposite). They produce large-scale standing waves (Rosby waves) within the upper atmosphere of the Northern Hemisphere. These waves occur in the lee of the western edge of the North American Cordilleras, to the east of the mountains of the European continent, and in the lee of the Tibetan Plateau. Without these mountains the pattern of atmospheric circulation would be much smoother. Consequently the Late Cenozoic uplift of these mountainous areas produced a series of waves and troughs within the pattern of atmospheric circulation of the Northern Hemisphere. There is therefore a strong north–south component to the circulation, which during a period of Milankovitch cooling would tend to draw down polar air into mid-latitudes, thereby intensifying the cooling episode in these areas. The occurrence and distribution of mountains within an 'Ice Age' may therefore have important consequences in determining its intensity.

Source: Ruddiman, W. F. & Kutzbach, J. E. 1990. Late Cenozoic plateau uplift and climatic change. *Transactions of the Royal Society of Edinburgh: Earth Sciences* **81**, 301–314. [Diagram reproduced by permission from: Boulton (1993). In: Duff (Ed.) *Holmes' Principles of Physical Geology*, Chapman & Hall, Fig. 21.12, pp. 450–451]

amounts of water. The increased cloud cover which would result from this would insulate the Earth, preventing heat loss and thereby cause a global warming. Although such ideas are currently fashionable, for the most part they remain untested.

2. . **Radiation variations.** A more plausible explanation for climate change involves variation in the amount of solar radiation received in the upper part of the atmosphere. Variation in the input of solar radiation to the Earth occurs at two broad scales. First, it varies at a high frequency due to Milankovitch radiation variations (Box 10.1). These determine the climatic pattern within an 'Ice Age'—the cyclic growth and decay of ice sheets—but cannot explain the onset of an 'Ice-house', since Milankovitch radiation variations occur continuously in both 'Ice-house' and

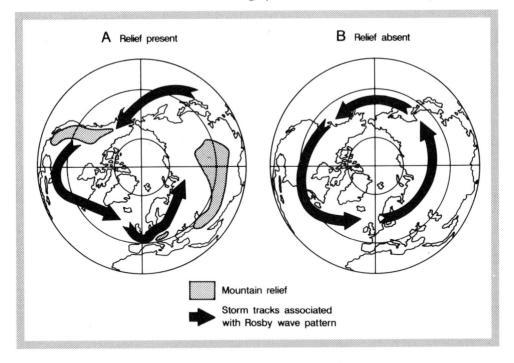

A Relief present

B Relief absent

Mountain relief

Storm tracks associated
with Rosby wave pattern

'Green-house' periods. The second scale of radiation variation is of much lower frequency, but higher magnitude. The solar system rotates around the centre of the galaxy once every 300 million years, a period known as the galactic year, and in so doing it passes through two stationary 'clouds' of hydrogen-rich particles which may cause fluctuation in the amount of solar radiation received by the Earth. Consequently, every 150 million years the Earth may receive less solar radiation. It has been suggested that the record of 'Ice Ages', or 'Ice-house' states, has an approximate periodicity of 150 million years. This is an interesting hypothesis, but at present the geological record of glaciation and palaeoclimate is not precise enough to allow any firm conclusions to be made.

3. **Variations in atmospheric carbon dioxide.** The budget of atmospheric carbon dioxide may help determine the Earth's climatic state. Carbon dioxide is a green-house gas and its proportion within the atmosphere will influence global temperatures: the more carbon dioxide, the greater the green-house effect and the warmer the climate. At the simplest level, atmospheric carbon dioxide concentrations are dependent on the rate of volcanic out-gassing, which is a function of plate tectonics, and on changes in the rate of continental weathering. Sea floor spreading, and the associated volcanic activity, provides the main input of carbon dioxide (volcanic gas) to the atmosphere, while chemical weathering of silicate rocks removes carbon dioxide. Carbon dioxide is removed by chemical weathering because rain dissolves carbon dioxide within the atmosphere to produce an acidic solution that breaks

down silicate rocks and locks the carbon molecules into the weathered rock. Periods of continental rifting and rapid sea floor spreading may therefore be associated with warmer climates, while periods of uplift and erosion, which reveal fresh silicate rocks for weathering, may have lower carbon dioxide contents and be cooler. The Cretaceous 'Green-house' period has been correlated with high carbon dioxide values produced by the rapid rate of continental rifting and sea floor spreading associated with the opening of the Atlantic Ocean. The high sea levels of this time would also have reduced the area of silicate rocks available for weathering. The atmospheric budget of carbon dioxide provides an important climatic forcing and may help to explain some of the variation between 'Ice-house' and 'Green-house' states recorded in the geological record. It is, however, important to note that the carbon dioxide budget is much more complex than the above discussion might suggest.

4. **The distribution of land and sea**. The nature of the Earth's surface may have an important control on climate, particularly through the relative distribution of land and sea (Box 10.4). Thus, there is a clear correlation between 'Ice-house' conditions ('Ice Ages') and the concentration of continental masses through plate tectonics in high latitudes. There are several reasons why this may happen. The presence of high-latitude land provides a surface for the accumulation of snow. Snow increases the reflectivity of land, its **albedo**, leading to an increase in the reflection of solar radiation back to the atmosphere, and as a consequence there is less radiation left to warm the land surface. Further, high-latitude land may block the poleward penetration of warm ocean currents and therefore limit the poleward transport of heat. 'Ice-house' conditions may therefore be explained by the concentration of continents in high-latitude areas through plate tectonics. In addition to helping to explain 'Ice-house' conditions, the distribution of continents may also explain 'Green-house' states. Most 'Green-house' periods are associated with large well-connected oceans, allowing the poleward transport of heat. For example, the mid-Cretaceous 'Green-house' world was one of the warmest periods in the Earth's history, and occurred at a time when ocean basins were well connected and the resulting ocean circulation pattern may have allowed efficient heat transfer towards the poles. Computer models of the Earth's atmospheric system have suggested that differences of between 3 and 4°C in the global average surface temperature may be induced by changing the distribution of continents on the surface of the Earth (Box 10.4). In addition, high sea levels and extensive continental flooding appear to be loosely correlated with periods of 'Green-house' conditions. Oceans and continental seas can absorb much more heat than land and also evaporate to give increased cloud cover, which helps retain heat. The high sea levels of the Cretaceous 'Green-house' phase may also, therefore, be of considerable importance in determining the pronounced warmth of this period.

The quest for an explanation of these large-scale climatic oscillations is a complex one. It is also unclear whether a single causal mechanism explains all events or if each episode was triggered by a different combination of factors. Despite this, however, changes in palaeogeography and in atmospheric carbon dioxide levels through time

BOX 10.4: THE ROLE OF CONTINENTAL DISTRIBUTION IN DETERMINING CLIMATE

Three-dimensional models of the climate system, known as **general circulation models,** have been used by Hay *et al.* (1990) to examine the role of continental distribution by latitude in determining climate. A series of experiments were conducted with a general circulation model in which the positions of the Earth's continents were changed. They examined the global climate produced by: (1) a large polar continent without an ice cap; (2) a large polar continent with an ice cap; and (3) a large tropical continent. The results, which are illustrated in the graph, clearly demonstrate the importance of continental distribution in determining climate. The experiment with a polar continent (without an ice sheet) produced a globally averaged surface temperature which was about 12°C cooler than when the model was run using a tropical continent. The presence or absence of an ice cap on the polar continent was also found to be very important in determining global temperature. The increased reflectivity of the ice cap enhances the rate at which the polar continent is cooled. These experiments help demonstrate the importance of the distribution of land and sea in determining the Earth's climatic state.

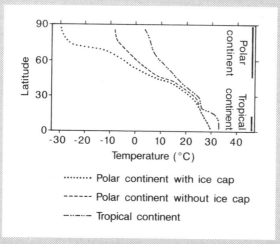

Source: Hay, W. W., Barron, E. J. & Thompson, S. L. 1990. Global atmospheric circulation experiments on an Earth with polar and tropical continents. *Journal of the Geological Society of London* **147**, 749–757. [Diagram modified from: Barron (1992) In: Brown *et al.* (Eds) *Understanding the Earth*, Cambridge University Press, Fig. 24.15, p. 502]

provide some of the most plausible explanations for the oscillation between 'Ice-house' and 'Green-house' climates in Earth history.

10.3 SEA LEVEL

Sea level has varied throughout geological time and is an important control on the type of sedimentary facies deposited. Not only does it determine the extent of marine sedimentation but also the depth of water and therefore the nature of the sediments (e.g. beach sands or offshore muds) deposited within a basin. The mechanisms by which global (**eustatic**) sea level have varied are poorly understood but the broad patterns of sea level variation appear to reflect changes in plate tectonic activity.

10.3.1 Sea Level Variation through Geological Time

Using the tools of facies analysis outlined in Section 5.1.2 one can obtain a broad picture of global sea level at a given point in time and space. By correlating the sea level records from different locations it is possible to recognise **eustatic** sea level events. These must be distinguished from local fluctuations in **relative sea level** caused by minor tectonic adjustments which alter land elevation. An estimate of global sea level variation throughout the last 500 million years has been established (Figure 10.3), but our understanding of Precambrian sea level is poor, since the sedimentary record from this period is fragmentary.

The Vail sea level curve in Figure 10.3, constructed in the manner discussed in Section 5.1.2, shows both first-order and second-order cycles of sea level variation. The first-order trend in sea level shows periods of high sea level during the Early Palaeozoic and the Cretaceous, separated by a period of much lower sea level. During the Cretaceous sea level may have been as much as 350 metres above its present level. Superimposed on this first-order pattern is a second-order sequence of major transgressions and regressions. About 14 second-order cycles can be identified, each of which lasted between 10 and 80 million years. During each of these second-order cycles the sea level rises, slows to a standstill and then drops rapidly. Superimposed on this pattern of second-order variation are numerous small-scale sea level fluctuations which probably reflect local rather than eustatic variation. These local variations may be caused by fluctuations in the size of ice sheets during an 'Ice Age', or by the build up of flexural stresses within moving plates which cause local tectonic adjustments which, in turn, cause variation in land elevation.

Why has sea level risen and fallen episodically throughout the Phanerozoic? Eustatic sea level changes can be brought about by either: (1) changing the volume of ocean water or (2) changing the volume of the ocean basins.

1. **Changing the volume of ocean water.** This can be achieved in three ways. First the volume of ocean water can be reduced by increasing the volume of land ice. In fact, sea level can fall by up to 150 metres simply through the growth of a large ice sheet.

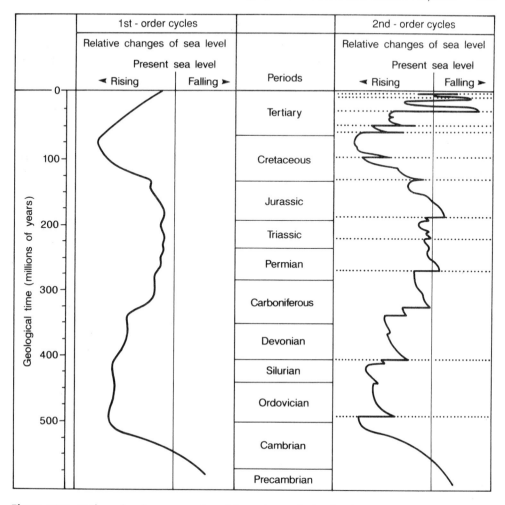

Figure 10.3 *Vail sea level curve produced from the analysis of sedimentary 'sequences' within seismic sections. [Modified from: Vail* et al. *(1977)* American Association of Petroleum Geologists Memoir 26, Fig. 1, p. 84]

This is very important in explaining the sea level oscillations experienced over the last two million years during the Quaternary 'Ice Age' and those of the Carboniferous 'Ice Age'. Second, the desiccation of shallow marine basins may reduce sea level. For example, during the Neogene the Mediterranean was an enclosed sea which was subject to periodic desiccation with consequential sea level changes of up to 15 metres. The final mechanism by which water volume can be changed is through temperature variation. For every degree Celsius that the temperature varies, the sea level can change by up to one metre, due to the thermal expansion or contraction of the water mass.

2. **Changes in ocean basin volume.** There are three ways in which the volume of the ocean basins may change. First, the mid-ocean ridges fill a significant part of the volume of an ocean basin. Sea floor spreading takes place at mid-ocean ridges, and variation in the rate of spreading can alter the volume of the ridge and therefore affect sea level. As new oceanic crust is formed and moves away from the ridge crest it cools, contracts and subsides. The new crust is hot and thermally buoyant and thus increases the topography and volume of the ridge. If the rate of sea floor spreading is rapid then new ocean crust is formed over a much larger area, and this means that the volume of the ridge is much greater than at ridges where sea floor spreading is slower. If the rate of sea floor spreading slows or decreases then the size of the ocean ridge would decrease, with the result that the volume of the ocean basin would increase, leading to a sea level fall (Figure 10.4). Conversely, an increased spreading rate would lead to a sea level rise (Figure 10.4). In this way, variations in the rate of sea floor spreading may cause global sea level to fluctuate by as much as 300 metres. Sea floor spreading rates do not appear to be uniform and their variation may account for many of the marine transgressions and regressions present within the stratigraphical record.

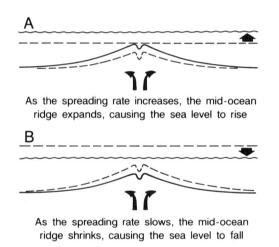

Figure 10.4 *The relationship of the rate of sea floor spreading to sea level fluctuations. [Reprinted with the permission of MacMillan College Publishing Company from: Lemon (1990) Principles of Stratigraphy, Merrill Publishing Company, Fig. 12.5, p. 306. Copyright © 1990 by MacMillan College Publishing Company, Inc.]*

A second mechanism leading to volume changes may be that of continental rifting and collision. During rifting, continental crust is stretched and continental area increased, a condition which may create marine transgressive episodes. Conversely, subduction along continental margins, or continental collision, decreases continental area and leads to marine regression. It has been suggested that the first-order pattern of sea level variation may be viewed in terms of the supercontinental cycle (Figure 10.5). The periods of high sea level during the Early

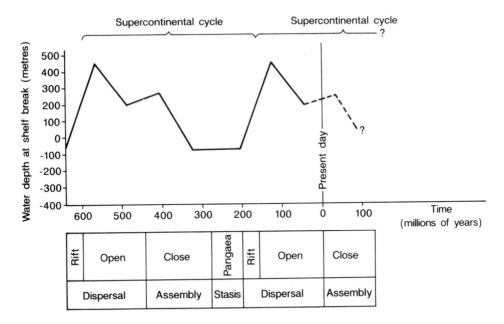

Figure 10.5 *The possible relationship between long-term eustatic sea level fluctuations and the supercontinental cycle. At times of continental assembly the average age of the ocean floor is at a maximum, spreading is slow, heat flow is low and therefore the ocean basins are deeper and eustatic sea levels lower. As a supercontinent splits up the average age of the ocean floor decreases, heat flow increases, spreading rates are high and, as a consequence, mean ocean depth decreases, causing high sea levels. [Modified from: Worsley et al. (1984)* Marine Geology *58, Fig. 13, p.391]*

Palaeozoic and the Cenozoic reflect periods of continental dispersion when sea floor spreading and continental rifting dominated, while the low sea levels of the Late Mesozoic reflect the formation of the supercontinent of Pangaea (Figure 10.5).

Finally, a third possible explanation for ocean volume changes is that of erosion and, more importantly, the resulting pattern of sedimentation. Increased sedimentation leading to the construction of large-scale features such as deltas could effectively displace the water mass of a basin and lead to a sea level rise. In practice, however, it is unlikely that this would have anything other than a regional effect in altering sea level.

Despite the range of possible explanations for eustatic sea level changes the pattern of first- and second-order variation recorded in Figure 10.3 cannot be completely accounted for. There is no single explanation for all sea level variations and it is possible that different mechanisms may have operated at different times.

10.4 BIOSPHERE

The **biosphere** encompasses all parts of the Earth in which life is present. In essence, it is equivalent to the sum total of life on Earth (the **biota**) at any given time. From the

Figure 10.6   *Coccolith limestone from the Upper Jurassic of southern England. Average coccolith diameter 5 μm. [Reproduced by permission of the Palaeontological Association from: Young & Bown (1991)* Palaeontology **34**, Pl. 1 p. 1]

first appearance of life on Earth, some 3500 million years ago, the individual components of the biosphere have evolved and the nature of the biosphere has changed. Thus,
the strata of different geological intervals each contain different assemblages of fossil
animals (fauna) and of plants (flora). Fossils give an insight into the nature of the
biosphere through geological time and, in some cases, are significant rock-forming
materials themselves (Figure 10.6). The following section outlines the broad evolutionary processes which have shaped the Earth's biosphere and then examines the broad
pattern of change through geological time.

10.4.1 Evolution and the Fossil Record

Evolution is the process by which species undergo gradual transformation through time. Although much of the modern study of evolution is tied to understanding the genetic processes of inheritance and species variation, it is only geology that can record the changes in evolving **lineages** (lines of descent of evolving organisms) through time. Geology is particularly important in helping us to determine the broadest scale of evolutionary changes in the development of the biosphere.

The changes in organisms up to and including the creation of species are encompassed within **microevolution**. It is the biological study of **genetics** that enables us to understand the mechanisms of microevolutionary change. The development of an individual organism is ultimately controlled by its **genes**. Genes are composed of **DNA** and are normally located in the cell nucleus arranged along elongate, thread-like bodies known as chromosomes. The DNA molecule carries a chemical code, known as the genetic code, which stores all of the information necessary for the function and development of an organism. As DNA is passed on from generation to generation, it provides the basis for the inheritance of characteristics and the continuity of life.

The **gene pool** is a collective name for all genes in a species or particular population. Evolution is expressed as a change in a population's gene pool through time. **Natural selection** is the process which operates on existing genetic variation (i.e. current species) to bring about successive changes in those species through time. Certain genes are lost from the gene pool because of the failure of individuals with those genes to reproduce and new genes are introduced through mutation. In this way natural selection takes place and the genetically determined characteristics that result from selection and which are in turn passed on through reproduction are known as **adaptations**.

New species arise from an existing species' gene pool when individuals form a population that diverges so much from other populations of the same species that the two groups can no longer interbreed: in other words when the interchange of genes between the two groups stops. This **reproductive isolation** is usually associated with geographical isolation of a population, or with an ecological restriction to a specific habitat. Put simply, if the individuals of one species are divided into two separate populations which become isolated from one another by a physical barrier then the flow of genes within the gene pool common to the two groups will be interrupted. In this case new species will eventually be created when individuals from the two groups have evolved so far that they can no longer interbreed. The creation of new species through such isolation is known to be extremely important in the development of new species.

In the past it was envisaged that evolution progresses gradually with many intermediate stages. This is a view which is very close to the ideas first expressed by Charles Darwin, the founder of modern evolutionary thought, in 1859. This model of evolution is often referred to as **phyletic gradualism**. In this model all the intermediate forms along an evolutionary path should be present; their absence in the geological record is assigned to the imperfection of the stratigraphical record. In recent years, however, scientists have questioned the traditional model of phyletic gradualism and have suggested that evolutionary change consists of periods of **stasis**, when no changes occur, punctuated by rapid periods of morphological development. Put another way this means that evolution is seen to occur in sudden bursts. This view of evolution has already been introduced in Section 4.1.3 (Box 4.3) and is known as **punctuated equilibria**. The fossil records of many groups often display periods of stasis and rapid change consistent with the theory of punctuated equilibria, although, in some cases, rapid and gradual change may be seen to occur together within a single fossil group (**punctuated gradualism**).

Species transitions of the sort described above are the domain of microevolution, but how do major new groups of animals and plants form? The geological record gives clear evidence of the diversification and development of new groups, and evolution at this scale is often referred to as **macroevolution**. The origin of major groups and the direct relationship of micro- and macroevolutionary processes are still an area of active research, but it is clear that the summation of microevolutionary processes must lead to large-scale changes in the biosphere. It is also clear that macroevolutionary processes may be a reflection of large-scale environmental changes: changes that result in the operation of natural selection processes that change the gene frequency in a large number of organisms at the same time. Two macroevolutionary processes that appear to be of extreme importance in the development of the biosphere are **mass extinctions** and **adaptive radiations**.

The fossil record provides a direct insight into the development of the biosphere through time. It is clear that there are periods when the standing diversity—the total number of different types of organism present on the Earth—has fallen dramatically over a short period of time. Species extinctions occur regularly in nature, but these **mass extinctions** involve numerous species and have had a dramatic effect on the evolution of the biosphere.

Mass extinctions appear to have occurred regularly throughout geological time. There have been at least five major mass extinctions in the history of life (Table 10.1) and several smaller ones (Figure 4.10). In fact, some authors have argued that mass extinctions may occur every 26 million years and are therefore a periodic phenomenon associated with a specific, possibly extra-terrestrial, cause (Box 10.5). This view has, however, received considerable criticism and is as yet unproven. The cause of mass extinctions is uncertain and there are four main hypotheses currently available to explain them.

1. **Sea level variations.** As we saw in the previous section sea level fluctuates through time. The effects of sea level change are known to have had a profound effect on the abundance and diversity of the marine fossil record. For example, the sea level rise at the beginning of the Cambrian is thought to have influenced the radiation of marine organisms by increasing the size and availability of suitable habitats (**ecospace**), that of shallow marine shelf areas. Conversely, a fall in sea level may increase the competition for ecospace amongst marine organisms and therefore lead to species extinction.

2. **Climatic change.** The Earth's climate has fluctuated between 'Ice-house' and 'Green-house' states through geological time. Climate, and in particular temperature, is an important limiting factor on organisms today. It is possible to identify

Table 10.1 *Major episodes of mass extinction in the Phanerozoic, together with candidate causes. Evidence for major meteorite impact is strongest only in the Late Cretaceous. [Percentage data from: Sepkoski (1982)* Geological Society of America Special Paper, ***190**, 283–289]*

Mass extinction event	Percentage marine families lost	Candidate causes
End Cretaceous	15%	Meteorite impact Volcanic winter
End Triassic	20%	Climatic effects (warming) Sea level fall
Late Devonian (Frasnian–Famennian)	21%	Climatic effects (cooling) Meteorite impact
End Ordovician	22%	Sea level fall Climatic effects (cooling)
End Permian	50%	Sea level fall Meteorite impact

extinctions that are probably climate controlled, such as the extinction of large mammals in the Pleistocene through the onset of the Quaternary 'Ice Age'. However, it is difficult to relate the mass extinction of both land and marine organisms directly to climate.

3. **Vulcanicity**. A major volcanic episode may have caused the end-Cretaceous mass extinction. The catastrophic eruption of volcanoes can lead to the input of large quantities of volcanic gases and ash into the upper atmosphere. This could produce a 'volcanic winter', a phenomenon which is caused when ash particles and aerosols are injected into the upper atmosphere. This reduces the amount of solar radiation received by the Earth's surface which in turn causes intense climatic cooling for a period of years. The end-Cretaceous extinction can be correlated with the eruption of large amounts of basaltic lava and it has been argued that these eruptions may have caused a short-term deterioration in the world's climate and thereby caused a failure of the global ecosystem.

4. **Extra-terrestrial impact.** The mass extinction at the Cretaceous–Palaeogene boundary (usually referred to as the Cretaceous–Tertiary or K–T boundary) is associated with a layer of clay rich in iridium (Figure 4.7). Iridium is commonly enriched only in meteorites and other extra-terrestrial bodies, and this has led to the suggestion that the mass extinction at the end of the Cretaceous may have been caused by the impact of an extra-terrestrial body. The impact of a meteorite would create what is known as an 'impact winter', similar to a 'volcanic winter' but induced by dust introduced into the atmosphere by the impact of a meteorite. It has also been argued that if the impact had taken place in an ocean, vaporisation of the water during collision would increase the global cloud cover leading to an accelerated green-house effect. Some authors have suggested that meteorite impacts occur at regular intervals as our solar system moves through the galaxy. These authors point to the apparent 26 million year cyclicity of extinctions identified within the geological record as evidence of this (Box 10.5).

BOX 10.5: PERIODICITY OF MASS EXTINCTIONS

Mass extinctions have long been recognised in the geological record, and most geologists accept that extinction events have occurred throughout the Phanerozoic. The work of J. J. Sepkoski and his co-worker D. M. Raup, however, suggests that the extinctions may actually occur at regularly spaced intervals. Sepkoski compiled data about the evolutionary development of marine animal families in the fossil record. Statistical analysis of these data indicates that the extinctions were periodic rather than random, and that the periodicity was regular in the Late Phanerozoic, from the Permian to the Recent, with an interval of 26 million years. Explanations for this periodicity have included regular, periodic impact of extra-terrestrial bodies with the Earth, regular volcanic eruptions, unspecified effects of magnetic reversals and sea level falls. However, no single cause can be invoked to explain all extinctions, and therefore there is little evidence to test the hypothesis. There are many criticisms of this work (e.g. Patterson & Smith 1987). Most importantly, as the data set was compiled from published records, it was a secondary data source, dependent on equal rigour of approach by the primary data collectors. In addition, many scientists reject the statistical techniques used as a proper test of periodicity.

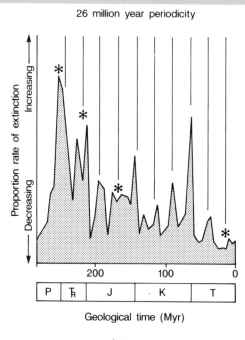

Sources: Raup, D. M. & Sepkoski, J. J. 1986. Periodic extinction of families and genera. *Science* **231**, 833–836. Patterson, C. & Smith, A. B. 1987. Is periodicity of extinction a taxonomic artefact? *Nature* **330**, 248–251. [Diagram modified from: McGhee (1989). In: Allen & Briggs (Eds) *Evolution and the Fossil Record*, Belhaven Press, Fig. 2.5, p. 36]

In summary, all four of these mechanisms may clearly be interrelated and it is unlikely that there is a single mechanism for all mass extinctions. As an example of this it is conceivable that a climatic change (cooling) would lead to a sea level change (fall) through the growth of ice sheets. Any resulting mass extinction could therefore be caused by a combination of both processes. Although the mechanism of mass extinction is still hotly debated, and is an area of popular interest, it is clear that mass extinctions have occurred and have had a profound effect on the evolution of the biosphere.

Periods of mass extinction are often followed and thereby balanced by periods of evolutionary expansion which effectively 'restock' the depleted biosphere. These expansions occur with the development of a successful group of organisms which proliferates and expands its geographical range rapidly, over a short space of geological time. The success of these organisms is usually associated with the development of a new adaptation which gives them an advantage in re-populating the environmental space vacated by the organisms killed off during the mass extinction. In other cases radiations may reflect other changes in the environment. For example, stromatolites only became common in the Proterozoic as oxygen increased (Figures 10.7 and 10.8). In some cases this may also be triggered by the utilisation of an old adaptation in a new way. The resulting evolutionary expansions are usually referred to as adaptive radiations.

Figure 10.7 *Photograph of Late Proterozoic stromatolites from Orkney, Scotland. Layers of sediment are trapped by algal mats, which give the irregular laminations and structures shown in the photograph. [Photograph: BGS]*

168

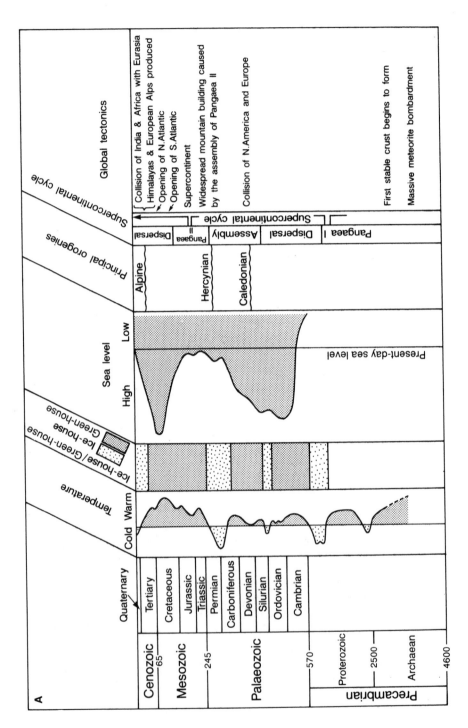

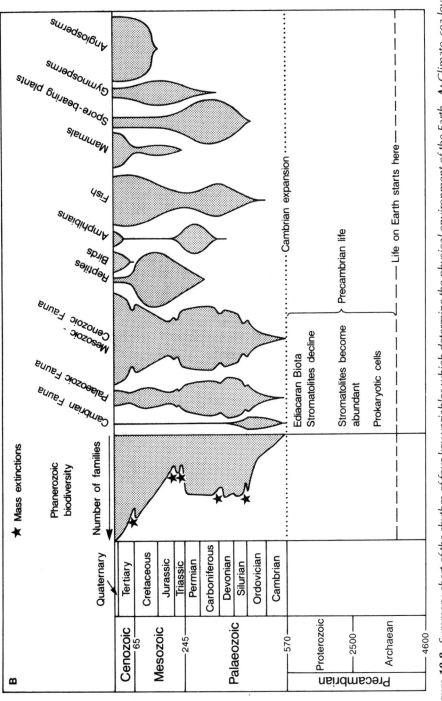

Figure 10.8 Summary chart of the rhythm of four key variables which determine the physical environment of the Earth. **A:** Climate, sea level and plate tectonics. **B:** The biosphere

Together, mass extinctions and adaptive radiations effectively control the stocking and restocking of the biosphere through time. This pattern of extinction and radiation is illustrated in Figure 10.8B, which shows the **clades** or clusters of evolving groups of animals and plants. These clusters represent the diversity of life on Earth at any given time, and they reflect the painstaking work of many specialists in compiling them. Extinctions are indicated by rapid reductions in diversity (i.e. the width) while radiations are shown by similar rapid increases in the faunal or floral diversity over a short interval of time (Figure 10.8B).

The Cambrian, Palaeozoic and Mesozoic–Cenozoic Faunas represent the major stocks of marine invertebrates. These faunas are often referred to as **evolutionary faunas**, and represent a similar pattern of diversification among a variety of different organisms. The Cambrian Fauna, the first shelly biota, was replaced after the Ordovician mass extinction by the Palaeozoic Fauna, dominated by brachiopod-rich assemblages, which in turn was replaced after the Permian mass extinction by the Mesozoic–Cenozoic Fauna, dominated by molluscs (Figure 10.8B). The evolution of vertebrates and of plant life is represented by clades in Figure 10.8B, and both show patterns of extinction and radiation. An example of an extinction event and its aftermath is illustrated by the extinction of the reptiles and the radiation of mammals at the end of the Cretaceous (Figure 10.8B). It is important to note, however, that adaptive radiations need not only be associated with extinctions. Particular examples of such radiations include the fish in the Devonian, and angiosperms in the Cretaceous. Both radiations are the result of the evolutionary success of a **key adaptation**, the development of shelly hard parts in marine invertebrates in the Devonian and the reproductive system (flowers and enclosed seeds) found in the angiosperms during the Cretaceous.

In the following section the record of change within the biosphere through geological time is discussed.

10.4.2 The Biosphere Through Geological Time

A record of the evolving biosphere is only possible through the presence of fossils within the geological record. The detail with which the biosphere can be reconstructed is usually limited by the quality with which the flora and fauna of a particular period are preserved. Fortunately, however, rare glimpses that approximate to the full diversity of life at specific points are given by the deposits known as **Lagerstätten**, a concept already introduced in Section 5.4 (Box 5.5). A Lagerstätte is a palaeontological 'bonanza' in which the conditions of preservation have often been so favourable as to preserve both skeletons/shells (hard parts) and soft tissue (soft parts) of organisms.

The development of the biosphere may be differentiated into two major phases: (1) the Precambrian, in which life first formed and diversified and (2) the Phanerozoic, in which rapid diversification of the biosphere took place creating the complexities of life that surround us today.

Life in the Precambrian

Two major phases can be identified in the development of life in the Precambrian: ancient life in the earliest or Archaean rocks and more complex life in the younger Proterozoic rocks.

The first fossil organisms are recorded from rocks in Western Australia which have been dated at some 3500 million years. These early cells, which are probably some form of bacteria, were able to survive in an anaerobic or oxygen-free environment, since it is thought that the Archaean atmosphere was composed of methane, carbon dioxide and hydrogen sulphide produced by volcanic degassing as the Earth's crust cooled. The energy source of these early organisms is the subject of some debate. The conventional view is that they depended upon preformed organic molecules to serve as an energy source. An alternative is that these early organisms were dependent on an external energy source, but not the Sun. This idea has developed through the recognition of present-day bacteria associated with volcanic hydrothermal vents on the ocean floor. Here organisms meet their energy needs through oxidising hydrogen sulphide in solution in the hot waters surrounding the vents, and this process is known as **chemosynthesis**.

In either hypothesis the evolutionary development of life is limited by resources, either by the amounts of preformed organic molecules, or by the distribution of suitable sites of inorganic energy. Therefore the appearance of a metabolism that utilised a truly renewable source of energy, such as light, was crucial to life on Earth. The development of **photosynthesis**, which uses light as an energy source, is one of the most significant events in the history of the biosphere. In photosynthesis solar energy is transformed to chemical energy when carbon dioxide is reduced by hydrogen, donated either by hydrogen sulphide in the case of certain anaerobic bacteria, or by water in the case of cyanobacteria and virtually all other photosynthetic organisms. Sulphur is released when hydrogen sulphide is used, while oxygen is released when water is the hydrogen donor. The first photosynthetic organisms lived in the early anaerobic atmosphere of the Earth. The development of photosynthesis is marked in the geological record by the occurrence of stromatolites (Figure 10.7). Stromatolites are formed by mats of cyanobacteria which trap sediment and thereby form mounds of sediment which are recorded in the sedimentary record. Their formation is closely associated with photosynthetic organisms. The development of a photosynthetic pathway which utilised water as the source of hydrogen and released oxygen as a by-product led to a gradual, but radical, change in the Earth's atmosphere. By the end of the Archaean the Earth's atmosphere was being enriched in oxygen by photosynthetic organisms.

In the Proterozoic the Earth's atmosphere had become increasingly oxygen-rich as a product of the cumulative effect of photosynthesis. In many ways the subdivision of the Precambrian into the Archaean and the Proterozoic approximates to the change from an anaerobic (oxygen-poor) to aerobic (oxygen-rich) atmosphere. Probably the most important evolutionary step during the Proterozoic was the appearance of the first **eukaryotic cell**. This type of cell not only has a nucleus and complex of chromosomes, but also a means of sexual reproduction and therefore a method of genetic variation and of natural selection. It is not known exactly when this event occurred, but it could not have taken place in the anaerobic environment of the Archaean, as eukaryotes require oxygen. From this type of cell multicellular organisms (**metazoans**) developed and these have been found in the geological record as far back as 1300 million years. The first evidence of tracks and trails left by metazoans are to be found in rocks 900 million years old. Following this in the Late Proterozoic is a distinct assemblage of multicellular organisms, known as the **Ediacaran Biota**, which was first found in the Ediacara Hills of Australia and dated at 600 million years. This biota and others like it (e.g. the **Charnian Biota** in England) is distinct from living organisms today, as although some members appear to resemble worms and jellyfish, others have a peculiar quilted structure which is unknown in animals today. This biota was extinguished at the end of the Proterozoic, apparently leaving no direct descendants.

Life in the Phanerozoic

The beginning of the Phanerozoic is marked by what is perhaps the most dramatic adaptive radiation, commonly referred to as the Cambrian expansion or explosion. This 'explosion' refers to the rapid diversification of animals with hard parts (i.e. shells and skeletons) and took place at the start of the Phanerozoic following the extinction of the Late Proterozoic Ediacaran Biota. This diversification may have been assisted by the effects of the widespread Late Proterozoic glaciation, which had two effects. First, it increased the mixing of water masses and therefore the supply of nutrients. Second, sea level increased significantly after the end of this 'Ice Age' and flooded large areas of continental margin creating shallow seas. The uptake of phosphate and then carbonate in these shallow seas by Cambrian organisms allowed them to form skeletal material for the first time.

In the early part of the Palaeozoic life was exclusively restricted to marine environments and was dominated by **trilobites**. Trilobites are related to arthropods, which are animals that have a segmented body, jointed limbs and are covered by a hard organic skeleton or shell (Figure 4.2). Crabs are typical arthropods. Trilobites dominated the marine environment in the Cambrian (the Cambrian Fauna), but at its end they suffered a mass extinction associated with a fall in eustatic sea level. Survivors of this event allowed a renewed radiation of the trilobites in the Ordovician, but for the rest of the Palaeozoic, and particularly following a second, major extinction at the end of the Ordovician, shelly fossils, particularly **brachiopods**, dominated the marine environment (the Palaeozoic Fauna). Animals with backbones (**vertebrates**) appeared in the Late Cambrian in the form of primitive fish, which continued to diversify throughout the later part of the Palaeozoic. The land was first colonised by arthropods during the Ordovician. This became possible through the diversification of land plants which acquired the adaptive advantage of a rigid stem and tissue which was capable of carrying nutrients throughout the plant (**vascular tissue**). A famous Lagerstätte, the Scottish Rhynie Chert, gives insight into the form of these early plants and more significantly of the land-dwelling animals which they supported during the Devonian. Continued evolution of the woody and root tissue of plants allowed the development of coal swamps in damp humid locations. These early plants were tied to aquatic habitats, because their reproductive system required water. The first land-dwelling vertebrates were amphibians derived from fish in the Early Carboniferous and they were closely followed by the first reptiles.

Three major extinctions took place in the Palaeozoic: (1) at the end of the Ordovician, which decimated marine faunas, (2) at the end of the Devonian and (3) at the end of the Permian, which effectively wiped out 50% of marine fauna (Table 10.1) and most of the early mammal-like reptilian fauna found on land. Each of these extinctions was matched by a subsequent adaptive radiation (Figure 10.8B).

The Mesozoic era started in the aftermath of the dramatic Permian extinction. The nature of the shallow marine fauna changed: the Palaeozoic Fauna, dominated by brachiopods, was replaced by a fauna dominated by molluscs, the Mesozoic–Cenozoic Fauna, still characteristic of the shallow marine environment today. Reef-building corals, like those of today, appeared for the first time and refilled the niche left by the extinction of the Palaeozoic corals. Cephalopods, hunting molluscs such as squid, became an important predator in the world's oceans. On land, the **gymnosperms** (naked seed plants) radiated widely, through the development of a reproductive system which was not dependent on water. Consequently this allowed the colonisation of dry lands. The flowering plants, **angiosperms** (enclosed seed) appeared at the end of the Mesozoic Era and are unusual because they resulted from an adaptive radiation which was not tied to a mass extinction. The angiosperms co-evolved with certain insects, which provided the means of fertilising the seed, and consequently the development of the angiosperms is matched by the diversification of insect life. The vertebrates continued to diversify, most notably the dinosaurs, and large reptiles dominated the land and sea. An extinction and subsequent radiation took place at the end of the Triassic, killing off much of the shallow marine fauna, as well as many of the early reptiles. The best documented and most controversial mass extinction took place at the end of the Cretaceous and led to the demise of the dinosaurs, the marine reptiles and many species of marine molluscs and plankton.

The Cenozoic Era followed this mass extinction and saw the development of the biosphere as we know it today. The most notable radiation, after the Cretaceous extinctions, was that of the mammals which restocked the niches vacated by the reptiles. In much the same way the birds filled the gap left by the extinction of flying reptiles. The angiosperms were largely unaffected by the Cretaceous extinctions and furnished the diversity of plant life we find today. The shallow marine sea-floor-dwelling (benthic) fauna was largely unaffected by the Cretaceous extinctions, but bony fish took over from the cephalopods as the dominant predator of the world's oceans. Many large mammals suffered extinctions towards the end of the Cenozoic which may reflect the onset of the Cenozoic 'Ice Age' or the first appearance of human hunters.

10.5 THE BIOSPHERE AS A CONTROL ON THE STRATIGRAPHICAL RECORD

The biosphere has contributed to the development of the Earth's stratigraphical record in two ways: first, through the modification of the local and even global environment and, second, through the production of primary geological materials.

The development of much of the Earth's present atmosphere was probably a direct result of the summation of the by-products of photosynthesis. Today, the atmosphere could be restocked with oxygen in just a few thousand years, through photosynthesis. In the Precambrian the development of the atmosphere took some thousands of millions of years. The oxygen-rich atmosphere obviously sustains growth but also has an effect on the nature of the minerals and sediments that are deposited. Redbeds are indicative of the oxidisation of iron minerals, and are common only from the Late Proterozoic onwards. Significantly, living organisms have an impact on the rate at which the land surface erodes and evolves. Prior to the colonisation of the land in the Early Palaeozoic there was little or nothing available to bind the surface of the soil and consequently the rates of erosion would have been high. With the evolution of land plants, rates of erosion and therefore of landscape change would have been reduced dramatically. Those areas where the vegetation cover was nearly complete would have experienced a relative fall in erosion and sediment production.

Living organisms also have a direct effect on the nature of sedimentation. Sea-floor and land-dwelling organisms disturb and mix accumulating sediment, while the bodies of such organisms enhance the rate of sedimentation at certain locations. In this way, the biosphere contributes directly to the stratigraphical record through the production of rocks of organic origin. The majority of carbonate rocks, for instance, are the product of the accumulation of vast numbers of shelly fossils, on either the microscopic or macroscopic scale (Figure 10.6). The Chalk in particular is a carbonate rock of organic origin which has developed through the accumulation of countless millions of the microscopic skeletal remains of planktonic marine plants. This deposit is widespread across Europe and America and is a significant part of the stratigraphical development of these regions. Reefs and other carbonate accumulations are also an important feature of the stratigraphical record. This type of organic deposit illustrates at one level how the biosphere has had a direct input into the development of the stratigraphical record.

10.6 SUMMARY OF KEY POINTS

We suggested at the start of this chapter that the stratigraphical record can be inter-preted in terms of four variables: plate tectonics, climate, sea level and the biosphere. These four variables combine to determine the nature of the rocks deposited or formed at any point on the surface of the Earth. The stratigraphical record is therefore a record of these four elements, which throughout Earth history have varied systematically. In this chapter we have explored the rhythms of change present within each of these variables and obtained an impression of the way in which the Earth has evolved over the last 4600 million years. There is much about this story that is incomplete or still unknown and our knowledge of the Precambrian is poor. It is also clear that these four strands of Earth history are interdependent and interact to a greater or lesser degree. For example, plate tectonics has an important influence on climate and sea level, which in turn has significantly contributed to the development of the biosphere and, feeding back, the biosphere has exerted its own influence in helping to moderate and evolve the Earth's climate (Figure 9.1).

The strands of this story of our planet have been brought together in Figure 10.8 in an attempt to summarise the key events of Earth history. This summation represents a key to what was happening to the Earth at any point in time and provides a model of the physical development of the Earth's crust, climate, sea level and biosphere. Our current understanding of Earth history has only been achieved through the efforts of numerous 'stratigraphical detectives' who have pieced together the evidence from across the world using the stratigraphical tool kit introduced in Part I.

If one was to take a time slice across this summary chart then the rocks formed or being deposited across the globe at that time would be the product of the subtle interplay of the global plate tectonic setting, sea level, climate and the evolving biosphere.

How well does the stratigraphical record of an area such as the British Isles fit within this global model? In the next chapter this question is explored and the British stratigraphical record is examined.

10.7 SUGGESTED READING

Many of the subjects covered in this chapter are introduced and discussed in the very readable book by van Andel (1985). The chapter by Fisher (1984) in *Catastrophes and Earth History* also provides a useful review of this subject. The detailed literature on the subjects reviewed in this chapter is vast and what follows is not intended to be a definitive guide, but simply to draw attention to some of the main sources of information.

Plate tectonics. Windley (1993) discusses the application of the plate tectonic model to the geological record with particular reference to the Precambrian. The distribution of the Earth's continents during the Palaeozoic is reported in Scotese and McKerrow (1990) and for the Mesozoic in Scotese (1991). The possible mechanisms underlying the supercontinental cycle are discussed at an introductory level by Nance *et al.* (1988) and in detail by Gurnis (1988).

Palaeoclimate. Frakes (1979) provides a good review of palaeoclimates through geological time, a story which is updated in Frakes *et al.* (1992). The chapter in Duff (1993) on palaeoclimate provides a good introduction to some of the possible mechanisms of climate change, particularly in the context of the Cenozoic 'Ice Age'. Barron (1992) provides a more sophisticated discussion of this subject. The work of Hay *et al.* (1990) gives an insight into the possible effects of continental distribution on the global climate, while the paper by Summerhayes (1990) introduces a series of papers in the *Journal of the Geological Society* on palaeoclimate (Volume **147**, pp. 315–392). The book by Imbrie and Imbrie (1979) is a good guide to Milankovitch radiation variations and their climatic impact.

Sea level. An accessible introduction to eustatic sea level variation in the geological record is provided by Lemon (1990). Hallam (1984) is a classic paper in this field and has much useful information, while Worsley *et al.* (1984) discuss the possible linkage between the supercontinental cycle and eustatic sea level. Hallam's textbook provides a readable overview of the subject (Hallam 1992).

Biosphere. Stanley (1986) provides a good general account of the development of the biosphere from a geological perspective, while in contrast Bradbury (1989) provides a biological view of the evolving biosphere. McAlester (1977) is a readable account, although a little dated. Both Allen & Briggs (1989) and Donovan (1989) contain papers which examine the development of life through mass extinctions and adaptive radiation. Finally, Briggs and Crowther (1989) is probably the most up-to-date source of information on the development of the biosphere through geological time.

Allen, K. C. & Briggs, D. E. G. 1989. *Evolution and the Fossil Record.* Belhaven Press, London.

Barron, E. J. 1992. Palaeoclimatology. In: Brown, G. C., Hawkesworth, C. J. & Wilson, R. C. L. (Eds) *Understanding the Earth.* Cambridge University Press, Cambridge, 485–505.

Bradbury, I. 1989. *The Biosphere.* Belhaven Press, London.

Briggs, D. E. G. & Crowther, P. 1989. *Palaeobiology: A Synthesis.* Blackwell, Oxford.

Donovan, S. K. 1989. *Mass Extinctions: Processes and Evidence.* Belhaven Press, London.

Duff, P. McL. D. (Ed.) 1993. *Holmes' Principles of Physical Geology.* (Fourth Edition) Chapman & Hall, London.

Fisher, A. G. 1984. The two Phanerozoic supercycles. In: Berggren, W. A. & Van Couvering, J. A. (Eds) *Catastrophes and Earth History.* Princeton University Press, Princeton, 129–150.

Frakes, L. A. 1979. *Climates throughout Geologic Time .* Elsevier, Amsterdam.

Frakes, L. A., Francis, J. E. & Syktus, J. I. 1992. *Climate Modes of the Phanerozoic.* Cambridge University Press, Cambridge.

Gurnis, M. 1988. Large-scale mantle convection and the aggregation and dispersal of supercontinents. *Nature* **332**, 695-699.

Hallam, A. 1984. Pre-Quaternary sea-level changes. *Annual Review of Earth and Planetary Sciences* **12**, 205–243.

Hallam, A. 1992. *Phanerozoic Sea-level Changes.* Columbia University Press, New York.

Hay, W. W., Barron, E. J. & Thompson, S. L. 1990. Global atmospheric circulation experiments on an Earth with polar and tropical continents. *Journal of the Geological Society of London* **147**, 749–757.

Imbrie, J. & Imbrie, K. P. 1979. *Ice Ages: Solving the Mystery.* Harvard University Press, Cambridge, Massachusetts.

Lemon, R. R. 1990. *Principles of Stratigraphy.* Merrill Publishing Company, Columbus, NY.

McAlester, A. L. 1977. *The History of Life.* (Second Edition) Prentice-Hall, Englewood Cliffs, New Jersey.

Nance, R. D., Worsley, T. R. & Moody, J. B. 1988. The supercontinental cycle. *Scientific American* **259**, 72–79.

Scotese, C. R. 1991. Jurassic and Cretaceous plate tectonic reconstructions. *Palaeogeography, Palaeoclimatology, Palaeoecology* **87**, 493–501.

Scotese, C. R. & McKerrow, W. S. 1990. Revised world maps and introduction. *Memoir of the Geological Society of London* **12**, 1–21.

Stanley, S. M. 1986. *Earth and Life through Time.* Freeman, New York.

Summerhayes, C. P. 1990. Introduction: palaeoclimates. *Journal of the Geological Society of London* **147**, 315–320.

van Andel, T. H. 1985. *New Views on an Old Planet.* Cambridge University Press, Cambridge.

Windley, B. F. 1993. Uniformitarianism today: plate tectonics is the key to the past. *Journal of the Geological Society of London* **150**, 7–19.

Worsley, T. R., Nance, D. & Moody, J. B. 1984. Global tectonics and eustacy for the past two billion years. *Marine Geology* **58**, 373–400.

11
The British Isles through Geological Time

The aim of this chapter is to illustrate how plate tectonics, climate, sea level and biosphere have interacted through geological time to create the pattern of Earth history which is preserved by the stratigraphical record. In this chapter we use one area of the Earth's crust as an example of how the pattern of temporal change within the four key variables determines the changing nature of the Earth's physical environment through geological time. Britain is our example, as its stratigraphical record is exceptionally complete: but equally we could have selected other parts of Europe, Asia, North America, or even Antarctica to illustrate the key elements of Earth history.

11.1 THE SCOPE OF THIS CHAPTER

In this chapter we tell the story of the geological development of the British Isles. We shall show how this story is the product of the interplay of plate tectonics, climate, sea level and the biosphere through time.

We shall, however, only discuss the key events in the geological development of the British Isles as they appear to us. These events are summarised at the ends of the appropriate sections. This chapter is therefore a personal view and, because of this, it does not represent a comprehensive treatise of British stratigraphy. Instead, it serves as an introduction to the global processes which have created and shaped one small part of the Earth. The text should be read with reference to the summary charts of the global record (Figure 10.8). Palaeogeographical maps are presented at the end of this chapter. These maps may be treated like a 'cartoon-strip' of the development of Britain through geological time. The present-day coastline of Britain is deliberately left off these maps, because Britain as a recognisable geographical entity has existed only in the most recent part of Earth history.

The subject of stratigraphy, like most other scientific subjects, is constantly changing as new ideas and theories are proposed, altering our perspective on Earth history. The story we give below is, therefore, the current view of the stratigraphy of the British Isles, a view which will no doubt become more precise as more evidence is obtained and as the interpretive framework of models and ideas is improved.

11.2 THE STORY OF BRITAIN

The story of Britain is one of continental construction through plate tectonics. This provides the framework for the development of the British stratigraphical record and, as a consequence, continental construction dominates any account of the early part of Britain's geological history. In turn, climate, sea level and the biota control the detailed characteristics of the sedimentary record and these are discussed throughout the text where appropriate. In addition, it is important to realise that other factors may overprint the clarity of this record. For example, due to plate tectonics the British Isles have gradually drifted northwards, from the Southern to the Northern hemispheres, through geological time (Figure 11.1). This northward drift has had an impact on the climate and biota of Britain, which overprints the effect of the Earth's global pattern. For example, as Britain travelled through the equatorial regions from the Southern to Northern hemispheres during the Carboniferous the globe as a whole experienced a period of 'Ice-house' conditions, yet in Britain conditions were tropical, due to its equatorial location.

We have divided the story of the British Isles into three blocks of time: (1) the Precambrian; (2) the Cambrian to Carboniferous, the period of the assembly of Pangaea; and (3) the Permian to Recent or 'the long quiet period'. The first of these blocks of time is vast, over 4000 million years, but in contrast to the other two we know very little about it and its stratigraphical record in Britain is relatively poor. The end of the Precambrian represents a major break within the stratigraphical record: a contrast between a primeval world with little biota and a world of abundant and diverse life. The rest of the geological record, about 600 million years, is best viewed in terms of two halves of the supercontinental cycle. The first half is the story of the assembly of Pangaea during the period from the Cambrian to the Carboniferous. This interval is one of dramatic events, of mountain building and upheaval, and as such is very distinct from the second half of the cycle which commenced with the start of the Permian. The second half is the story of the large Pangaea supercontinent and its dispersal. This time interval has often been referred to as 'the long quiet period', since apart from regular transgressions and regressions of the sea and frequent episodes of crustal rifting, it is relatively uneventful.

11.3 THE MYSTERY OF THE PRECAMBRIAN

The Precambrian represents an immense interval of geological time about which we know relatively little. The Precambrian record of the British Isles is fragmentary and

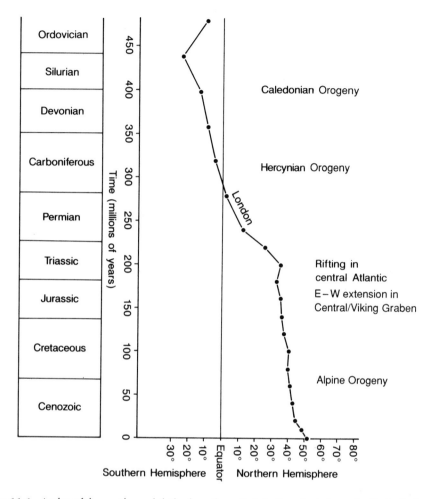

Figure 11.1 *A plot of the northward drift of southern Britain (London) since the Early Ordovician.* [Modified from: Glennie (1990). In: Glennie (Ed.) Introduction to the Petroleum Geology of the North Sea, Blackwell, Fig. 2.7, p. 43]

because it has been subject to geological processes operating over a great interval of time, it is often difficult to piece together as a coherent story. What is clear, however, is that during the Precambrian and Early Palaeozoic the British Isles did not form part of a single continent. If we examine the combined geological record of Scotland and Northern Ireland and compare it with that of southern Britain it is clear that they are dissimilar, each having stronger affinities with other continental masses. In this way we can deduce that Scotland and parts of Northern Ireland were attached to the North American continent, while southern Britain was probably attached to Africa. These two halves of the British Isles did not come together until the Caledonian Orogeny in the middle of the Palaeozoic. As a consequence it is misleading to try to reconstruct the geography of the area of the British Isles during this period, as the two halves were separated by a large extent of ocean. Therefore, in any discussion of the Precambrian of the British Isles, Scotland and parts of Northern Ireland must be considered separately from southern Britain and Ireland.

11.3.1 Scotland and Northern Ireland

Within Scotland one can recognise a number of distinct Precambrian terranes, each with its own stratigraphical record and each representing a different 'snap-shot' of Precambrian environments. These terranes were only brought together during the Caledonian Orogeny. The relationship between the different terranes during the Precambrian is unclear, although research continues in an attempt to unravel the complexities of this story. We can recognise three Precambrian terranes in Scotland and Northern Ireland (Figure 11.2) and these are: the Hebridean Terrane, Northern Highland Terrane and Grampian (Dalradian) Terrane.

1. **Hebridean Terrane.** The Hebridean Terrane consists of two distinct stratigraphical elements of very different age: (1) Lewisian basement and (2) Torridonian.

    The **Lewisian** consists of high-grade, regionally metamorphosed rocks, mainly gneisses. These rocks are thought to represent a fragment of Archaean granitic crust which may have formed around 2900 million years ago and which has been subsequently heavily metamorphosed. Perhaps surprisingly the Lewisian also contains rocks in which sedimentary structures can still be identified. This shows that the original rocks prior to metamorphism were not simply composed of granitic crustal rocks but also some sedimentary rocks. The Lewisian records several phases of metamorphism. The first phase of metamorphism probably took place around 2700 million years ago and produced a diverse assemblage of metamorphic rocks. A second phase of metamorphism has been dated to 2400 million years, and this re-metamorphosed the already heavily altered rocks. The Lewisian has been successfully divided chronologically on the basis of these two phases of metamorphism, which were punctuated by the intrusion of a series of north-west to south-east trending dykes, known as the Scourie Dykes. The story of this discovery has already been introduced in Section 7.1.1. The Scourie Dykes are unmetamorphosed by the early phases of metamorphism, but were deformed and altered by the later episode of metamorphism. On this basis two episodes of metamorphism and orogenesis have been recognised: (1) Scourian Orogenesis (pre-dyke emplacement) and (2) the Laxfordian (post-dyke emplacement).

    Throughout the period of the formation of the Lewisian rocks, Scotland was an integral part of the North American continental margin and the nature of the Lewisian basement has strong parallels with that found in central Greenland. The Lewisian crystalline rocks were part of a much larger area of ancient Precambrian crystalline rocks known as a **shield**. Precambrian shields are found at the heart of

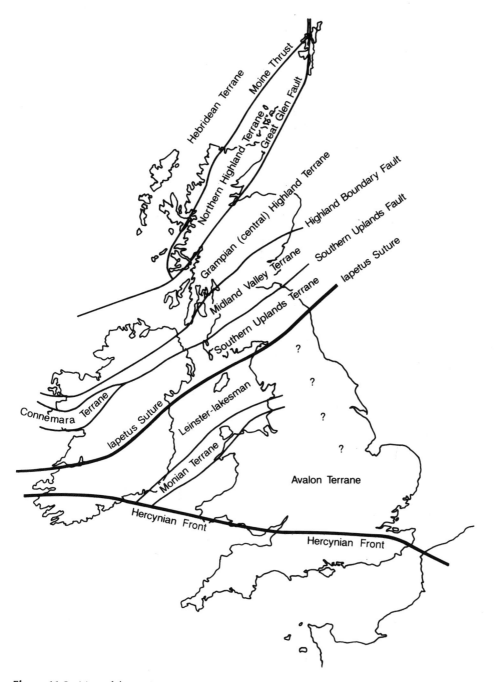

Figure 11.2 *Map of the main tectonic terranes recognised within the British Isles. [Modified from: Bluck et al. (1992). In: Cope et al. (Eds) Atlas of Palaeogeography and Lithofacies, Geological Society Memoir No. 13, Fig. 1, p. 3]*

most of the present-day continents. Unfortunately, although we know sedimentary rocks were deposited upon this early crust we have little evidence of the nature of the developing sedimentary record in the Scottish Lewisian. Some observations can, however, be made from the few sedimentary structures preserved, which show ripple marks indicative of a shallow marine environment. The intensity of metamorphism has been such that no evidence of early life has yet been recovered from these metasedimentary rocks. The development of the Archaean biosphere is therefore unrecorded in Scotland. Evidence from rocks of similar age in Africa and Australia do show, however, that prokaryotic life was developing in an oxygen-poor atmosphere at this time.

The Lewisian basement is overlain unconformably by much younger Torridonian sediments which are of Proterozoic age. These sediments were deposited by huge braided rivers flowing from the Archaean basement of Greenland some 900 to 600 million years ago. The basal part of the Torridonian sequence, the Stoer Group, is composed of coarse clastic sediments which buried the eroded landscape of the Lewisian basement and infilled a series of rift basins. Study of the sedimentary structures preserved in these rocks confirms the flow direction of rivers was from the North American shield. These sedimentary rocks are mostly red in colour and this is consistent with an interpretation of their deposition in a hot, arid climate. The presence of microfossils within these rocks gives some evidence of life within these environments. Subsidence of the rift basins or the evolution of new rifts caused the Stoer Group to be faulted, tilted, eroded and submerged by a shallow sea. The basal units of the younger Torridon Group were deposited by rivers which prograded into this shallow sea. These units contain simple, marine microfossils. Subsequently, deposition became dominated by braided rivers and by large alluvial fans emerging from a highland region to the north-west, located somewhere over Greenland.

2. **Northern Highland Terrane.** This consists of the Moine rocks of the North West Highlands. Although these rocks have been heavily deformed and metamorphosed several times, we have sufficient evidence to suggest that they were originally deposited in a shallow marine clastic environment. The age of Moine sedimentation is poorly constrained, but it has been frequently suggested that they may represent an offshore facies of the Torridonian and therefore be of equivalent age. Radiometric dating illustrates that this is improbable because some dates for the metamorphism of the Moine show that this took place during the Grenvillian Orogeny, an event which predates the oldest ages attributed to the Torridonian sediments. At least two distinct phases of deformation can be identified within the Moine. In the first phase, the original sediments were deformed and metamorphosed during the Grenvillian Orogeny, the effects of which can also be identified in Scandinavia and in North America. The origin of the Grenvillian Orogeny is unclear, but it may have resulted from the closure of an ocean or the collision of an island arc with a continental margin. The second phase of deformation occurred during the Caledonian Orogeny in the Early Palaeozoic. In the course of the Grenvillian Orogeny the Moine rocks were thrust over the older Hebridean Terrane along a series of thrust planes of which the largest is called the Moine Thrust. These

Caledonian?

thrusts resulted in complex structural relationships which are clearly displayed in the Assynt area of north-west Scotland today.

3. **The Grampian (Dalradian) Terrane.** The rocks of the Grampian Terrane consist of highly deformed sediments known collectively as the Dalradian, which were laid down in a variety of depositional environments over a period of 150 million years. For the most part the Dalradian sediments were deposited in marine conditions within large fault-controlled basins associated with a period of crustal stretching along the North American plate margin. The onset of Dalradian sedimentation was probably marked by the transgression of a Late Proterozoic sea across a complex basement consisting of Moine, Torridonian and Lewisian rocks. Dalradian sedimentation postdates the Grenvillian Orogeny, which was responsible for the initial deformation of the Moine.

The Dalradian preserves the record of the Late Proterozoic 'Ice-house' as it contains a succession of tillites. These tillites are the product of the widespread glaciation which can be recognised in many parts of the world at the close of the Proterozoic. At Port Askaig, on the island of Islay, glacio-marine deposits are recorded, and these equate with the widespread evidence of glaciation found over many of the Late Proterozoic continents. The Proterozoic biosphere was dominated by the development of stromatolites, which indicate an increase in the level of oxygen in the atmosphere (Figures 10.7 and 10.8). The lower part of the Dalradian sequence contains the remains of numerous stromatolites. The transition to the Phanerozoic marine biosphere dominated by shelly faunas is indicated in the upper part of the sequence, above the tillites, which contains a sparse fauna consistent with that associated with the Early Cambrian faunal expansion (Figure 10.8).

The crustal extension which provided the basins for Dalradian sedimentation changed to compression in the later Proterozoic and this caused the initial deformation and metamorphism of these sediments. This period of deformation, which occurred in several phases between the Late Proterozoic and the Early Ordovician, is usually referred to as the Grampian Orogeny, which is seen by some as the initial phase of the Caledonian Orogeny. The effects of the Grampian Orogeny were felt along much of the eastern margin of the North American plate and may have been caused by the collision of an island arc.

The palaeogeographical relationship of each of these terranes during their formation is unclear, despite many years of research. No simple picture of Scotland's Precambrian history is currently available due to the fragmentary nature of the record.

11.3.2 Southern Britain

The Precambrian basement of southern Britain is much younger than that found in Scotland, dating back only to the Late Proterozoic (*c.* <700 million years). Proterozoic rocks are poorly exposed in England and Wales and small, but important, outcrops occur in areas of central England, the Welsh Borderland, the Malvern Hills, South Wales and Anglesey. The Precambrian basement beneath southern Britain is widely believed to have developed through the gradual accretion of island arcs, associated accretionary prisms and fore-arc basin sediments on to the edge of the Proterozoic supercontinent of Gondwana, probably in the vicinity of Africa. Of particular importance is the occurrence of soft-bodied multiple-celled organisms (metazoans) within the Late Proterozoic rocks of the Midlands and Wales. These organisms, known as the Charnian Biota, after Charnwood Forest in the English Midlands, were amongst the first Proterozoic metazoans to be discovered. They are of particular importance because they can be used as guide fossils for the Late Proterozoic, due to their widespread distribution and overall similarity to the Australian Ediacaran Biota. These biotas comprise a variety of soft-bodied organisms which were widespread in the latter part of the Proterozoic, and are indicative of the increased oxygen content of the atmosphere. The soft-bodied organisms that comprise them left no descendants to survive into the Phanerozoic, suffering an extinction some time before the close of the Proterozoic.

KEY EVENTS: PRECAMBRIAN OF BRITAIN

Plate tectonics: Very little is known about the global plate tectonic setting during the Precambrian, although it has been suggested by many that there was at least one supercontinental cycle. The Precambrian terranes of northern Britain contain evidence of several distinct phases of orogenesis. The cause of these orogenic episodes is unclear. The Precambrian basement of southern Britain is believed to have been formed by the accretion of a series of island arcs onto the margin of the Gondwana continent. The precise details are at present still a matter of some debate.

Climate: Our knowledge of the climate of the Precambrian is poor. We do know that the Proterozoic was mostly warm and arid, and this is confirmed by the Torridonian redbeds. The Late Proterozoic, however, was an 'ice-house' interval and the Scottish Dalradian sequence preserves evidence of this 'Ice Age' in the form of tillites. There is little evidence in southern Britain of global climatic changes in the Proterozoic.

Sea level: Our knowledge of eustatic sea level changes in the Precambrian is poor. Evidence of fluctuation of sea level in the Archaean is suggested by the preservation of sedimentary structures in the Lewisian. Definite eustatic episodes in the Proterozoic are illustrated by the flooding of the Lewisian basement to produce the Stoer Group sediments of the Torridonian, and the transgression at the base of the Dalradian sequence.

Biosphere: The Archaean biosphere was mostly composed of single-celled prokaryotes. The Proterozoic was dominated by stromatolites and, much later, by early metazoans. The Lewisian has not yet yielded evidence of Archaean life. Prokaryotic cells and stromatolites have been recovered from Scottish Proterozoic sequences. The Proterozoic multicellular organisms are recorded in the Charnian Biota of southern Britain.

11.4 CAMBRIAN TO CARBONIFEROUS: THE CONSTRUCTION OF PANGAEA

The rocks of the Palaeozoic contain the record of the assembly and construction of a large supercontinent, **Pangaea**. This construction took place through two large cumulative orogenies, the **Caledonian** in the Early Palaeozoic and the **Hercynian (Variscan)** in the Late Palaeozoic, both of which had a significant effect on the sedimentary record of Europe and North America. The geographical area we know as the British Isles was extensively affected by both orogenies and Britain as we know it today preserves a remarkable geological record of the assembly of Pangaea.

The Caledonian Orogeny culminated in the continental collision of three separate components: **Laurasia**, the early North American continent of which Scotland formed part; **Baltica**, the northern European/Scandinavian continent; and **Avalonia**, a microcontinent composed of southern England and Newfoundland. The record of collision is extremely complex because most of its components were directly affected by the deformation created by the orogeny. Its most obvious effects are the development of the Caledonian mountain chain. These mountains comprise the highland areas of Britain, Scandinavia and the eastern seaboard of North America. More significantly, the orogeny led to the creation of a large northern continent that we call **Laurussia.**

The Cambrian period was one of great environmental change (Figure 10.8). The climate improved, changing from 'Ice-house' to 'Green-house' states, and eustatic sea level rose. As a consequence, in many areas Cambrian marine sediments overstep the Precambrian basement. The rise in sea level, which followed the Late Proterozoic glaciation, flooded the continental margins to create large shelf seas in which a shelly biota, the Cambrian Fauna, could evolve. This fauna had its forerunners in a range of small shelly fossils which replaced the soft-bodied Charnian Biota. The Cambrian faunal development or expansion is often referred to as an explosion; an example of an adaptive radiation. In addition to the opening of new habitats, the radiation may have been a response to changes in deep ocean circulation, induced by the switch from 'Ice-house' to 'Green-house' states. An abundance of deposits rich in phosphates across the world at this time suggests that there was strong upwelling of deep ocean water, rich in nutrients, during the Early Cambrian. This upwelling may have provided the phosphatic material which was then incorporated into the shells of the first shelly organisms of the Phanerozoic. The rise in sea level and the appearance of the shelly Cambrian Fauna can be recognised along both the northern (Scotland) and southern (England) margins of the ocean which separated them, and are indicative of the pattern of global environmental change at this time.

From the Late Proterozoic onwards the continental components later involved in the Caledonian Orogeny were separated by the **Iapetus Ocean**. The existence of the Iapetus Ocean during the Early Palaeozoic was first recognised using fossil evidence. The Cambrian Fauna is dominated by trilobites which appear to have been confined to the relatively shallow levels of the newly flooded continental shelves. Trilobites were unable to cross the deep Iapetus Ocean, and consequently the trilobite fauna typical of Laurasia was very different from that found in the shelf seas of Baltica and Avalonia. We can plot the closure of the Iapetus through time, as these two different faunas

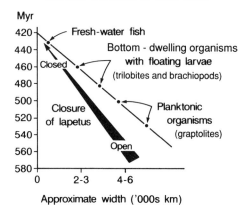

Figure 11.3 *A graph illustrating the progressive closure of the Iapetus Ocean based on faunal evidence. In the early stages of the ocean the only animals that were common to both margins were planktonic organisms (graptolites). With progressive closure of the ocean, a common fauna of bottom-dwelling marine animals with free-floating larvae existed on both margins. Closure of the ocean is indicated by the presence of the same type of fresh-water fish either side of the Iapetus Suture. [Modified from: McKerrow & Cocks (1976)* Nature *263, Fig. 1, p. 305]*

became mixed, through the later part of the Early Palaeozoic, as the ocean closed (Figure 11.3). As northern and southern Britain were divided by the Iapetus Ocean the Early Palaeozoic is perhaps best viewed as the evolution of two separate ocean margins, which came together during the Caledonian Orogeny (Figures 11.4 and 11.5).

11.4.1 The Development of the Northern Iapetus Margin

The evolution of the northern margin of the Iapetus was very complex involving the accretion or welding of a number of different terranes on to the plate margins. The situation is further complicated by the fact that these terranes are **allochthonous**, a term which refers to the fact that they have been moved from their original place of formation. Movement of the terranes was lateral, between the colliding plates of Laurasia, Baltica and Avalonia, and took place during the final stages of the closure of the Iapetus Ocean. In outline, the evolution of the northern margin appears to consist of three main events prior to the closure of the Iapetus. These events were:

Figure 11.4 *A series of sketch maps of Britain and North West Europe during the Palaeozoic.* **A, B** *and* **C:** *The closure of the Iapetus Ocean and the Caledonian Orogeny.* **D:** *One interpretation of the Hercynian Orogeny. [Based on information in: McKerrow (1988). In: Harris & Fettes (Eds)* The Caledonian–Appalachian Orogen, *Geological Society Special Publication No. 38, pp. 405–412; Glennie (1990). In: Glennie (Ed.)* Introduction to the Petroleum Geology of the North Sea, *Blackwell, pp. 34–77; Scotese & McKerrow (1990)* Memoir of the Geological Society of London *No. 12, pp. 1–21; Berthelsen (1992). In: Blundell et al. (Eds)* A Continent Revealed: The European Geotraverse, *Cambridge University Press, pp. 11–32]*

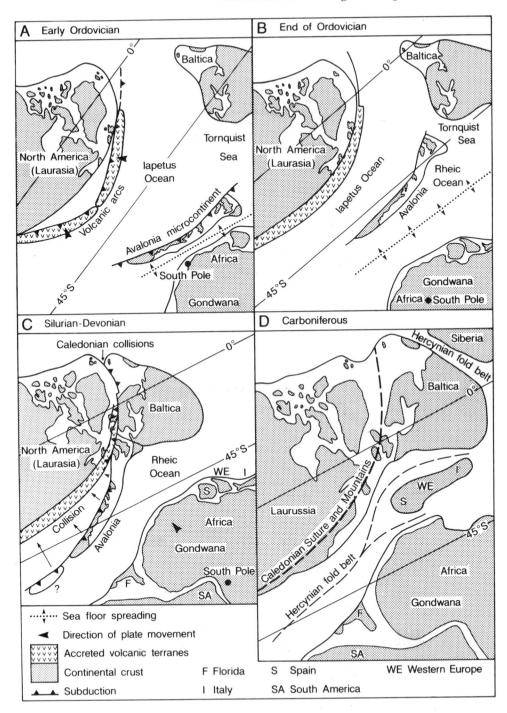

A   Early Ordovician

Baltica

North America (Laurasia)

Iapetus Ocean

Tornquist Sea

Volcanic arcs

Avalonia microcontinent

Africa

South Pole

Gondwana

0°

45°S

B   End of Ordovician

Baltica

North America (Laurasia)

Tornquist Sea

Iapetus Ocean

Rheic Ocean

Avalonia

Gondwana

Africa ● South Pole

0°

45°S

C   Silurian-Devonian

Caledonian collisions

Baltica

North America (Laurasia)

Rheic Ocean

Collision

Avalonia

WE   I

S

Africa

Gondwana

South Pole

F

SA

?

0°

45°S

D   Carboniferous

Hercynian fold belt

Siberia

Baltica

Caledonian Suture and Mountains

Laurussia

WE

S

I

Hercynian fold belt

Africa

Gondwana

F

SA

0°

45°S

⋯⋯ Sea floor spreading

◄ Direction of plate movement

vvv Accreted volcanic terranes

Continental crust

▲▲ Subduction

F Florida

I Italy

S Spain

SA South America

WE Western Europe

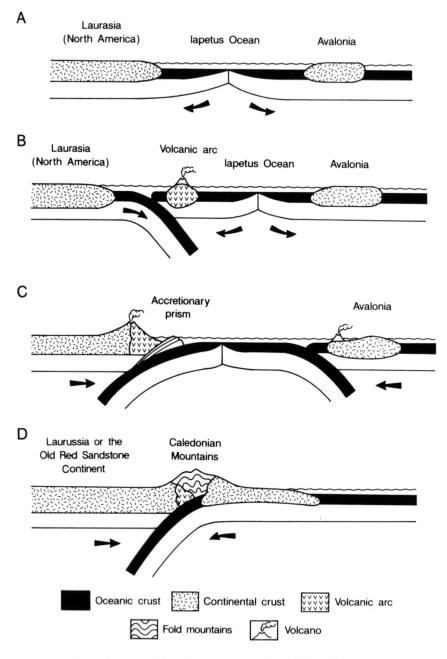

Figure 11.5 *Cartoon of the plate tectonic setting for the Caledonian orogeny*

1. An initial orogenic phase. This event, called the Grampian Orogeny, took place some time prior to the Ordovician and is thought to have been caused by the collision of an island arc with the eastern margin of Laurasia. This orogeny was responsible for the first phase of deformation and metamorphism of the Dalradian rocks of the Grampian Highlands. These rocks were again deformed and metamorphosed during the Caledonian Orogeny.
2. Terrane accretion. Following the initial orogenic phase, further accretion took place of island arcs, back-arc basins, accretionary prisms and fragments of ocean crust known as **ophiolites** and consisting of ocean floor basalts, fragments of the mantle rocks and oceanic sediments. This was probably associated with a reversal in the direction of subduction, with the Iapetus Ocean crust descending northwards beneath Laurasia.
3. Terrane displacement. On the closure of the Iapetus Ocean the accreted terranes were displaced laterally, due to the oblique collision of Avalonia with Laurasia, which took place during the Silurian.

As a consequence of these complex events the history of the northern margin is difficult to interpret. Clearly, the juxtaposed terranes we observe today did not necessarily form next to one another and are therefore stratigraphically distinct. Because of this it is more appropriate to discuss the development of the northern margin with reference to individual terranes. Five distinct terranes can be recognised (Figure 11.2), Southern Uplands, Midland Valley, Grampian (Dalradian), Northern Highland and Hebridean, and these are separated by the Southern Uplands Fault, Highland Boundary Fault, Great Glen Fault and the Moine Thrust, respectively.

As we have seen above, the Hebridean and Northern Highland terranes are composed of ancient Archaean (Lewisian) and Proterozoic (Moine) rocks which formed part of the leading edge of the North American plate. The Grampian, Midland Valley and Southern Uplands terranes all evolved along different parts of the active plate margin to the east of the North American plate. These terranes were juxtaposed by lateral movement along the axis of the suture created by the Caledonian Orogeny.

The Grampian Terrane consists of Dalradian sediments that were deposited in a marine shelf setting on the leading edge of the Laurasian plate. These sediments were first deformed and metamorphosed during the Grampian Orogeny. This deformation may have resulted from the collision of an island arc with the plate margin, although no evidence of an island arc has yet been found within the Scottish sector of the former plate margin (Figures 11.4 and 11.5). The Midland Valley Terrane is composed of sediments deposited in fore- and back-arc basins as well as the volcanic products of the island arc itself. The Southern Uplands Terrane consists of a stacked sequence of sediments of Ordovician and Silurian age which were deposited in a deep-water, marine setting. This sequence has been interpreted as an accretionary prism which accumulated along the subduction zone caused by the descent of the Iapetus Ocean beneath the Laurasian plate. The sedimentary facies of continental slope sands and deep water muds is consistent with such a trench environment. However, it has been suggested that a similar sequence of sediments could be produced in what is termed a **successor basin**. A successor basin is produced along the line of suture between two continental masses. In this case, the successor basin would represent the final remnant of the Iapetus Ocean, although not now floored by ocean crust. Continued compression of the basin by the continents could produce structures reminiscent of an accretionary prism and similar to those found in the Southern Uplands. Upon the closure of the Iapetus all three terranes (Grampian, Midland Valley and Southern Uplands) were moved laterally to their present positions (Figure 11.6). At the same time crustal compression and shortening within the leading edge of the Laurasian plate brought the Hebridean and Highland terranes into close contact as the Highland Terrane was thrust over the Hebridean terrane along the Moine Thrust.

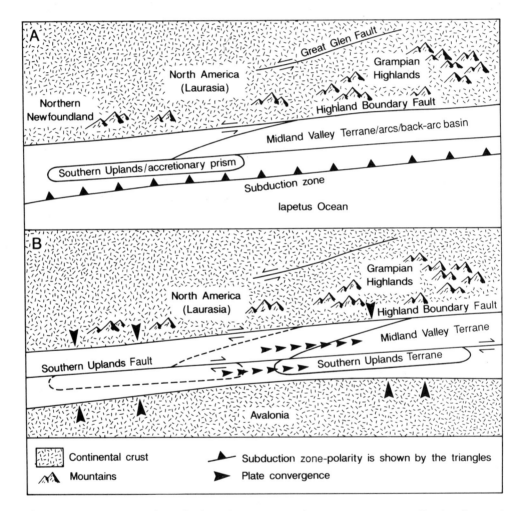

Figure 11.6 *Cartoon to show the lateral movement of tectonic terranes (Midland Valley and Southern Upland terranes) along the Caledonian Suture*

11.4.2 Development of the Southern Iapetus Margin

During the Early Palaeozoic, southern Britain and Newfoundland were part of a microcontinent, **Avalonia**, which rifted away from the southern continent of Gondwana and drifted northwards to form the southern margin of the Iapetus Ocean (Figure 11.4). The original location of Avalonia on the margin of Gondwana and the precise timing of the rifting are unknown. However, it is clear that Avalonia effectively formed the southern margin of the Iapetus from the Cambrian onwards.

The widespread marine transgression which followed the initiation of the 'Green-house' state in the Cambrian is recorded along this margin and the sediments deposited are rich in trilobites (Cambrian Fauna). Marine sediments were deposited in a deep sedimentary basin over much of Wales. Shallower shelf deposits were laid down in the English Midlands. Subduction of the Iapetus Ocean beneath the margin of Avalonia commenced in earnest during the Ordovician. This led to the creation of volcanic island arcs, in a marine setting, in both Wales (Snowdonia) and northern England (Lake District) (Figures 11.4 and 11.5). Subduction ceased for a time in the Late Ordovician and the margin became quiet. Marine deposition occurred in a variety of settings in both shallow and deep water conditions during this period. The end of the Ordovician is delimited by a worldwide mass extinction event (Figures 4.10 and 10.8). This event is recorded in a reduction of the marine fauna present within these sediments. The extinction was coincident with a return to 'Ice-house' conditions during the Ordovician and a reduction in sea level as a polar ice cap grew on the Gondwana continent. Other important developments in the biosphere also occurred in the Ordovician; in particular the invasion of the land surface by plants (Figure 10.8). This invasion is suggested by microfossils which give evidence of plant spores in Wales. Plant macrofossils, the remains of plants themselves, are found in Silurian sediments. This marks a major change in the biosphere as plant life, and subsequently the animal life associated with it, developed on land in the later Palaeozoic.

The Silurian is marked by a sea level rise and a gradual return to a 'Green-house' world (Figure 10.8). The 'Green-house' conditions promoted the growth and development of shelf carbonates, particularly the development of reefs of coral and algae. Brachiopod-rich faunas replaced trilobites as the dominant marine fauna, and are found in a range of marine shelf facies deposited on the Avalonian margin at this time. This pattern of deposition was interrupted towards the end of the Silurian with the closure of the Iapetus Ocean.

11.4.3 Northern and Southern Iapetus Margins United: The Birth of Britain

With the closure of the Iapetus the continents of Laurasia, Avalonia and Baltica were joined and finally by the Devonian the two parts of Britain were united. The

Caledonian Orogeny culminated at this time with continued compression and lateral displacement along the axis of the suture. These effects moved southern Britain into line with Scotland during the Devonian (Figure 11.6 and Map 2). The final phase of lateral displacement occurred in the mid-Devonian when movement along the Great Glen Fault brought the Northern Highlands into line with the Grampian Mountains. Rapid uplift of the Caledonian Mountains and the emplacement of granite batholiths in northern Britain complete the story of the Caledonian Orogeny and the unification of the two parts of Britain.

At the close of the Caledonian Orogeny, Britain was part of a large continent, Laurussia, often referred to as the Old Red Sandstone Continent. The continent is given its informal name because of the nature of the sediments which were deposited over much of its geographical area in the Devonian. Thus Britain was the site of the deposition of a post-orogenic sediment, known as the Old Red Sandstone, which was being eroded from the newly emergent mountains (Map 2). The Old Red Sandstone Continent experienced a continental semi-arid climate consistent with the 'Green-house' world of the Silurian, and this continued into the Devonian, although climate became more continental as a consequence of the closure of the Iapetus. Continentality was enhanced by a eustatic sea level fall (Figure 10.8). Climate and continentality influenced the deposition of coarse, clastic, predominantly red sediments in a variety of continental settings. The deposition of this continental facies actually commenced in the Late Silurian and, by the Devonian, at the culmination of the Caledonian Orogeny, almost all of Britain was land, except for south-west England. In northern Britain, close to the continental interior, coarse sandstones were deposited in basins between mountain blocks (intermontane basins). Drainage in this region was often internal (i.e. not open to a sea), flowing into basins and large lakes developed between mountains. In southern Britain an alluvial plain was crossed by large braided and meandering river systems. The sedimentary record from these areas contains abundant evidence of vigorous water flow and flooding. In some regions, ephemeral lakes and pools formed between the main river course during floods and were quickly desiccated. In south-west England the alluvial plain passed into marine conditions. This part of Britain formed the northern margin of a broad back-arc basin formed by plate convergence to the south of Britain, in northern France (Map 2). Deltaic sediments and carbonate reefs were deposited along the northern margin of this basin.

Evidence of the Devonian biosphere is patchy in Britain. Fresh-water fish radiated and were widespread in the ephemeral lakes and streams of northern Britain at this time. The first evidence of a complex land-based ecosystem of spore-bearing plants (*Rhynia* and *Asteroxylon*) and land-dwelling animals is recognised in northern Britain at Rhynie. The Devonian limestones of south-west England preserve evidence of reef faunas on the periphery of the Old Red Sandstone Continent. A global mass extinction event has been recognised in the Frasnian–Famennian (end Devonian) (Figures 4.10 and 10.8), but although this had a dramatic effect on the global biosphere, there is little evidence of its effects in Britain, apart from extinctions of some of the reef-building organisms in south-west England.

KEY EVENTS: THE CLOSURE OF IAPETUS AND THE BIRTH OF BRITAIN

Plate tectonics: During the early part of the Palaeozoic the Earth's continents began to move together in the construction of the supercontinent of Pangaea. This resulted in a major orogenic event, known as the Caledonian Orogeny. Northern and southern Britain, which had been separated by the Iapetus Ocean, were united, as the Iapetus Ocean closed as part of this overall continental aggradation.

Climate: Throughout this period, global climate was in a 'Green-house' state, with the exception of a brief period of 'Ice-house' conditions during the Ordovician. Although polar ice caps grew in Gondwana, located over the South Pole, the effects of this global cooling were not pronounced in Britain, which was experiencing deep water sedimentation. The return to 'Green-house' conditions following the Ordovician 'Ice-house' event is recorded in Britain by the deposition of terrestial redbeds during the Devonian.

Sea level: From the Cambrian onwards eustatic sea level experienced a rapid rise which peaked in the Early Ordovician when the growth of the polar ice caps resulted in a sea level fall. Sea level continued to rise during the Silurian, with the return to 'Green-house' conditions. The general sea level rise during this period resulted in the deposition of shallow marine sediments over much of southern Britain. After the Silurian, sea level fell, and this is consistent with the mostly terrestrial sediments of Britain at this time.

Biosphere: The start of the Cambrian saw a rapid radiation in shelly marine organisms, known as the Cambrian explosion or expansion. At the end of the Ordovician a major extinction event decimated the Cambrian Fauna, which was replaced by a brachiopod-rich fauna, the Palaeozoic Fauna. The Cambrian Fauna and succeeding Palaeozoic Fauna are represented in the sedimentary sequences of the northern and southern ocean margins. The invasion of the land took place in the Ordovician; this is recorded in the sedimentary sequences of Wales and Scotland. The Late Devonian extinction event is largely unrecorded in Britain because of the nature of the terrestrial sedimentation.

11.4.4 The Final Assembly of Pangaea

The **Hercynian Orogeny** led to the final assembly of Pangaea (Figure 11.4). The orogeny commenced in the Devonian, but in Britain its effects were felt throughout the Carboniferous. The orogeny involved the collision of the southern continent of Gondwana (Africa, South America, Antarctica, Australia) with the newly created Old Red Sandstone Continent (Laurussia), between which a number of smaller microcontinents were sandwiched. The line of suture between these continental masses lay well south of Britain, along a line that now runs through central France. In front of the advancing continent of Gondwana, a marginal sea or back-arc basin opened during the Early Devonian. This basin, known as the **Rheno-Hercynian Basin**, opened as a result of crustal rifting behind an island-arc subduction complex in southern Brittany (Figure 11.7). Floored by ocean crust, this

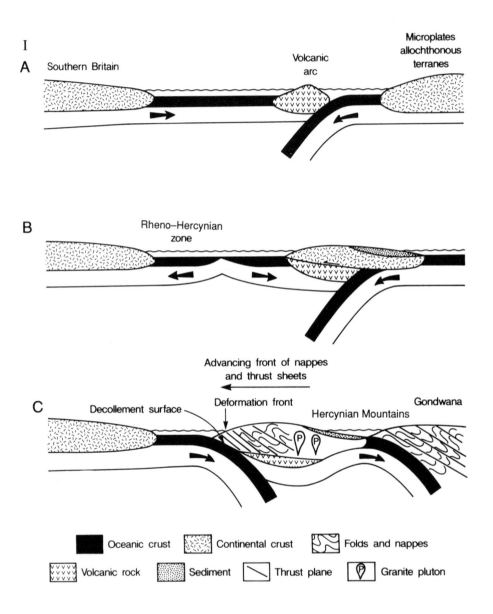

I

A   Southern Britain

Volcanic
arc

Microplates
allochthonous
terranes

B   Rheno–Hercynian
zone

Advancing front of nappes
and thrust sheets

C   Decollement surface   Deformation front   Gondwana

Hercynian Mountains

Oceanic crust   Continental crust   Folds and nappes

Volcanic rock   Sediment   Thrust plane   Granite pluton

II

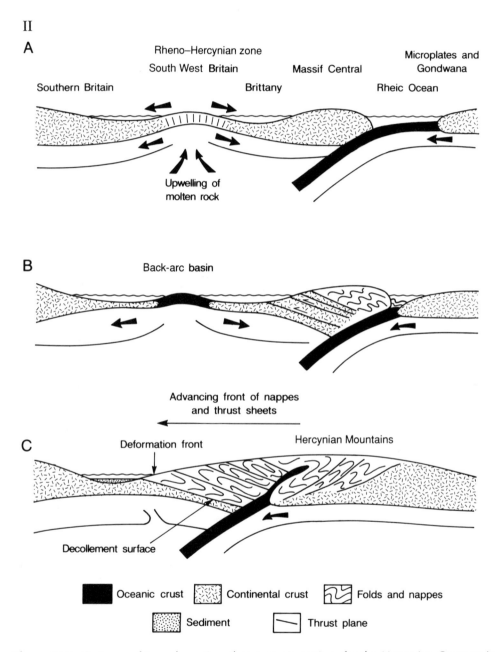

A

Rheno–Hercynian zone

South West Britain   Massif Central   Microplates and
                                                    Gondwana

Southern Britain   Brittany   Rheic Ocean

Upwelling of
molten rock

B   Back-arc basin

Advancing front of nappes
and thrust sheets

C   Deformation front   Hercynian Mountains

Decollement surface

Oceanic crust   Continental crust   Folds and nappes

Sediment   Thrust plane

Figure 11.7 *Cartoons of two alternative plate tectonic settings for the Hercynian Orogeny in southern Britain. [Based on information in: Berthelsen (1992). In: Blundell et al. (Eds) A Continent Revealed: The European Geotraverse, Cambridge University Press, pp. 11–32; Dunning (1992). In: Duff & Smith (Eds) Geology of England and Wales, The Geological Society of London, pp. 523–561]*

basin was slowly consumed as compression caused by the collision of Gondwana with Laurussia forced deformation northwards.

The style of deformation in the Hercynian Orogeny was very different from that experienced in the earlier Caledonian Orogeny. The Hercynian Orogeny was mostly completed through **thin-skin tectonics**. This type of deformation occurs when the upper layers of the crust are uncoupled from rocks deeper in the crust along a **decollement** surface (Figure 11.7). This surface acts rather like a polished floor underneath a rug: the rug is easily 'rucked up' in a series of big folds when it is moved quickly over the shiny surface. In southern Britain the rug is represented by a series of giant overfolds, called nappes, and large sheets of rock known as thrust sheets which advanced northwards during the orogeny. The advancing 'front' of deformation finally stopped along a line that runs from South Wales to Kent. This line, known as the **Hercynian Front**, denotes the limit of the thin-skin style of deformation, although the effect of the orogeny was felt in other ways north of the line. During the Hercynian, ocean crust from the floor of the back-arc basin was thrust into the pile of deformed sediments. This ocean floor is recorded as an ophiolite complex now exposed in the Lizard Peninsula in south-west England. The final phase of the Hercynian Orogeny was the intrusion of a large granite batholith which cuts across the complexly folded Hercynian sequence. Erosion of these Hercynian uplands has unroofed this batholith to reveal parts of it in the granite upland moors, of south-west England, such as Dartmoor and Exmoor.

While the southern margin of Britain and Europe was being compressed and deformed by continental collision, northern Britain was experiencing a period of crustal extension associated with the upwelling of molten rock. A consequence of this is that the later sedimentation of the Carboniferous in northern Britain is associated with a block and basin topography produced by a series of rift basins and horst blocks. Crustal extension took place along pre-existing fractures, formed during the Caledonian Orogeny, as well as along new fractures created during the Hercynian Orogeny.

The Old Red Sandstone Continent of Britain was flooded from the south for the first time in the Early Carboniferous (Map 3). Eustatic sea level curves show an upturn which equates with a rise in sea level at this time. This is a reverse of the fall experienced during the Devonian which led to the flooding of much of the low-lying areas of England and Ireland. The Early Carboniferous continental sea was warm and tropical, ideal for the deposition of coral reefs and shelly limestones, still often dominated by brachiopod faunas. The Carboniferous Limestone facies is persistent over much of Britain and its equivalent is found in other parts of Pangaea. In Britain, the pattern of deposition within this shallow shelf sea was strongly controlled by the pattern of blocks and basins created by the rifting.

Towards the middle part of the Carboniferous the global climate reversed into an 'Ice-house' state (Figure 10.8). This created a global fall in sea level as water was locked into the ice sheets which advanced over much of Gondwana. Evidence of this glaciation is limited in Britain, because despite global refrigeration, Britain was still experiencing tropical humid conditions, caused by Britain's location close to the equator at this time

(Figure 11.1). Direct evidence of this can be found in the presence of **bauxite** deposits in Scotland dating from this time interval, indicating warm, damp, equatorial conditions.

Deposition of the Early Carboniferous carbonate rocks was periodically interrupted in northern Britain by prograding deltas and there was a regular interchange between carbonate and clastic sediments, known as the **Yoredale Cycles**. This pattern can be seen to be a regular alternation or cycle of carbonate and deltaic clastic sediments. Patterns of cyclic sedimentation are commonly referred to as **cyclothems** and were probably caused in this case by the local tectonic adjustment of horst blocks within the block and basin topography. In this way uplift of the blocks would cause local regressions and the progradation of deltas, depressions of the block would lead to local transgressions and a return to carbonate deposition. A major eustatic sea level fall did, however, take place in the Middle Carboniferous. During this interval, deltaic and fluvial river systems prograded southwards over the carbonates. This regression and return to clastic sedimentation was caused by renewed uplift of the Caledonian Mountains, either as a result of the release of intraplate stress generated during the Caledonian Orogeny, or as a result of the compressional regime of the developing Hercynian Orogeny to the south (Map 4). The result of the regression was that much of northern Britain was covered by coarse clastic sediments known as the **Millstone Grit** (Map 4). Extremely thick deposits of Millstone Grit (up to 2 km) were deposited in the Pennine regions. These gritstones are periodically interrupted by **marine bands** formed by short-lived transgressions (marine incursions) and these effectively provide event horizons with which to correlate the sequence. The marine incursions can be related to eustatic sea level fluctuations because of their widespread occurrence and because of this they may be related to fluctuations in the size of the Gondwana ice sheet. The number of fluctuations appears to have decreased through time and the intensity of fluvial deposition also appears to have declined, such that in the Late Carboniferous there was a marked increase in tropical vegetation on the tops of the deltas and along the river floodplains (Map 5).

The forest vegetation of the Late Carboniferous consisted for the most part of **lycopod** spore-bearing plants, which underwent a major radiation in the Late Devonian and Carboniferous (Figures 3.5B and 10.8). Although restricted to swampy environments because of their reproductive cycle, these plants were able to build large trees for the first time, with the development of vascular woody tissue. On their death, these lycopods and other spore-bearing plants accumulated as peat in the humid swamps of this large deltaic region. Compression of this peat has produced the coal on which the industrial wealth of Britain was originally built. It also provides the source material for much of the natural gas of the southern North Sea. Rare, uncompressed **coal balls** formed by the accumulation of carbonate in pockets, thereby preventing compression, produce a Lagerstätte which provides an important glimpse into the detail of the lycopod structure. The coal is found in association with lacustrine muds and coarse clastic sediments introduced by local fluctuations in the sea level. Periodically minor marine transgressive episodes record the presence of continued sea level fluctuations, but local tectonic movements appear to be the principal control on sedimentation within the coal swamps.

KEY EVENTS: THE FINAL ASSEMBLY OF PANGAEA

Plate tectonics: The process which started with the Caledonian Orogeny was completed by the collision of Gondwana with Laurussia to form a large super-continent (Pangaea). This collision resulted in the Hercynian Orogeny which had its greatest effect in the southernmost portion of Britain.

Climate: Global climate was in a 'Green-house' state which was reversed towards 'Ice-house' during the Carboniferous, when polar ice sheets again developed over Gondwana. This interval of global cooling is not well recorded in Britain, due to its equatorial position at this time.

Sea level: At the start of the Carboniferous, eustatic sea level rose, a trend which was reversed with the growth of polar ice sheets in the latter part of the Carboniferous.

Biosphere: The early part of the Carboniferous saw the development of widespread carbonates consistent with 'Green-house' conditions and high eustatic sea level. The Carboniferous interval saw the radiation of tree-forming lycopods. These formed in the swamp conditions created by the sea level fall in the later Carbo-niferous. Both elements are well represented in the British Isles.

11.5 THE PERMIAN TO RECENT: THE LONG QUIET PERIOD

The Permian to Recent interval followed the assembly of the supercontinent of Pan-gaea, and many of the events associated with the early part of the interval are the after-effects of its construction. The stratigraphical record of the later part of this interval, however, reflects the development and opening of the Atlantic as well as the minor tectonic impact (in Britain) caused by the closure of the Tethys Ocean of southern Europe (Figure 10.1).

By the end of the Carboniferous, the area we know as Britain was amalgamated into the supercontinent of Pangaea. Compressive stresses from the Hercynian Orogeny led to the gradual inversion of the Carboniferous basins, such that areas which were formerly depositional basins now became gentle rolling uplands available for denu-dation. A new tectonic style was also developing. Extensional tectonics, on a regional scale, dominated the Permian throughout the British Isles and the North Sea region. These tensional stresses within the crust resulted in a series of rift basins, particularly in the western England, Irish Sea and North Sea regions. In the North Sea, two major rift basins, the **Southern North Sea Basin** and the **Northern North Sea Basin** were created, separated by a raised block known as the **Ringkøbing-Fyn High** which was later cut by two north–south trending grabens (**Central** and **Viking grabens**). There are two possible explanations for these tensional stresses which led to the Permian extensional tectonics: either they were caused by the release of intraplate stress generated during the Hercynian Orogeny, or alternatively they may have been caused by a forerunner of the tensional regime which was to open the Atlantic in the Cretaceous. Whatever the cause of these stresses, the resulting graben and half-graben basins and rift valleys across Britain produced the depositional setting in which the post-orogenic mollasse sediments of the Hercynian Orogeny were deposited.

The Permian depositional environments were largely determined by Britain's position in the centre of the Pangaea supercontinent, and by the global climatic state (Map 6). Britain was just north of the equator and during the Permian the global climate switched from an 'Ice-house' to a 'Green-house' state, a state which was to exist for the remainder of the Palaeozoic and Mesozoic (Figure 10.8). For some 30 to 40 million years Britain was part of a desert at the heart of Pangaea which was comparable in size to that of the Sahara today. The Early Permian saw extensive desert weathering of the Carboniferous basement and the deposition of sediment in dunefields. Large playas or desert lakes existed between the intensely weathered Hercynian uplands and were frequently desiccated to give salt pans (Map 6). Desert conditions continued throughout the Permian, but during the Late Permian a localised marine incursion of the Boreal Ocean gave rise to the Bakevellia and Zechstein Seas to the west and east of England, respectively (Map 7). Within these two warm, shallow, saline seas carbonate-rich rocks were deposited. The shorelines of these two seas were subject to periodic movement and the deposition of carbonate rocks was interrupted by evaporites and clastic sediments. In fact, the sediments along the margins of these seas have a cyclic nature, the cycles recording the alternation between carbonate, evaporite and clastic sediments. These local sea level fluctuations probably reflect local tectonic movements and are not believed to be eustatic in nature. The close of the Permian is associated with the most dramatic mass extinction in Earth history, in which 50% of marine families became extinct (Figures 4.10 and 10.8). This event marks one of the major chronostratigraphical boundaries, that of the Palaeozoic–Mesozoic boundary, but marine sequences preserving a record of this boundary are not found in Britain. This extinction may have been caused by a dramatic fall in sea level which reduced the area of shelf sea and increased the continentality of Pangaea (Figure 10.8).

Following the major marine regression in the Permian, desert conditions reasserted themselves during the Triassic (Map 8). The tensional stresses that had started in the Permian continued, and led to the development of north–south trending graben structures, such as the Worcestershire Graben. Triassic sedimentation is largely confined to these grabens and half-grabens formed between the Hercynian and Caledonian uplands. The Early Triassic sequence in Britain is almost completely continental and devoid of marine influence. The climate was hot and arid, although intense but infrequent rainfall ensured that coarse clastic sediments accumulated in alluvial fans and braided rivers. Evaporite deposits were abundant and formed in large lakes (playas) which frequently dried out to give salt pans. Reptilian remains in the form of trace fossil footprints complete the pattern. There is, however, some evidence of a brief marine incursion during the mid-Triassic, which led to the deposition of red dolomitic mudstones.

The youngest Triassic rocks record the start of a gradual marine transgression from the south caused by rising sea level in the Tethys Ocean. These deposits consist of fine-grained clastic and carbonate sediments containing a limited marine fauna. The presence of stromatolites within these deposits suggests that this advancing sea was tidal, since stromatolites prefer tidal conditions. The marine transgression at the close of the Triassic continued on and off throughout the rest of the Mesozoic and culminated in the Late Cretaceous high-stand when vast areas of the world's continents were flooded. The end of the Triassic, like the end of the Permian, is associated with a major extinction in which about 20% of all animal families became extinct (Figures 4.10 and 10.8). This extinction is again difficult to recognise in Britain due to the absence of stable marine environments suitable for fossil preservation.

The eustatic sea level rise which started in the Late Triassic culminated in the Jurassic with a flooding episode that effectively drowned much of the north-west European area of Pangaea (Map 9). Relict Caledonian and Hercynian upland areas, including much of Scotland, North Wales, Cornwall and the Anglo-Brabant Landmass, formed a series of islands within this epicontinental sea. Throughout much of the Jurassic sea level continued to rise. In the Early Jurassic mudrocks were deposited uniformly over much of the flooded area by pelagic sedimentation (Map 9). Despite this, in locally shallow areas of the sea, limestones and ironstones were deposited (Box 3.4). Correlation of the marine sediments of the Jurassic is excellent because of the abundance and suitability of ammonites as guide fossils within them. The marine sediments also document the dominance of molluscs, which replaced the brachiopods as the dominant marine invertebrates, following the mass extinction event at the end of the Permian (Figures 4.10 and 10.8).

The progressive rise in sea level during the Jurassic was interrupted by a minor regressive phase in the mid-Jurassic (Map 10). This was coincident with domal upwarping in the North Sea region and with local volcanism. Both phenomena were probably the products of crustal stretching caused by the initiation of a tensional regime, which was to lead ultimately to the opening of the Atlantic Ocean. This upwarp exposed the Lower Jurassic sediments of the North Sea to erosion and was followed by controlled subsidence in this area throughout the rest of the Jurassic and Cretaceous. Coarse clastic sediments were deposited in some parts of northern Britain, particularly in north-west Scotland. Here, river deltas coalesced to form an alluvial plain, and similar environments existed in the Viking Graben of the North Sea.

Away from the North Sea region the epicontinental sea decreased in depth during this phase of upwarping and several new islands emerged. This shallowing of the warm continental sea led to a migration of limestone deposition northwards during the mid-Jurassic. This trend was, however, not to last, as towards the end of this interval renewed transgression and sea floor subsidence gave rise to the continued deposition of mudrocks. The Late Jurassic is thus associated with a continued increase in sea level and the extensive deposition of mudrocks, particularly the **Kimmeridge Clay** (Map 11). This organic-rich clay facies is similar to that deposited in the Early Jurassic, and both sets of mudrocks provide the major source rocks for North Sea oil. This pattern is, however, complicated by continued and increased uplift and rifting, referred to as the **Cimmerian Movements**, within the North Sea region. By the end of the Jurassic sea level began to fall, which had the effect of isolating southern England from the depositional environments of northern Britain and the North Sea where the deposition of mudrocks continued. In the south, the Wessex Basin was an area of carbonate deposition, which gradually transformed into a hypersaline lagoon in which coniferous forests developed in a Mediterranean-type climate.

Marine regression continued at the start of the Cretaceous and the crustal rifting associated with the opening of the Atlantic Ocean led to uplift within Britain (Figure 11.8:

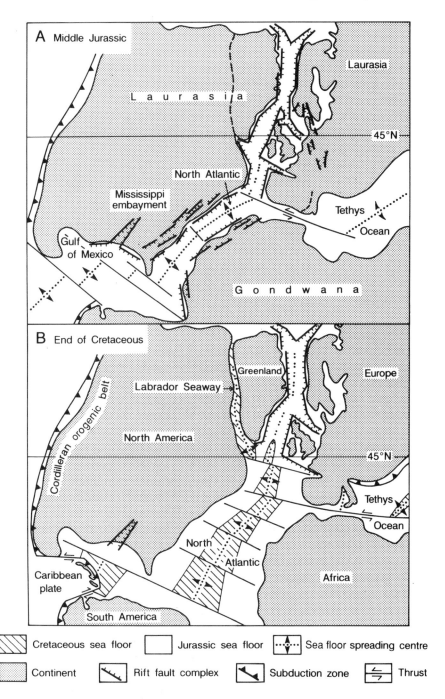

Figure 11.8 *A series of sketch maps which show the opening of the North Atlantic. [Modified from: Dott & Batten (1988)* The Evolution of the Earth, *McGraw-Hill, Figs 16.2 and 16.3, pp. 463 and 464]*

Map 12). At this time Britain was mostly land and marine deposition was confined to north-east England and to the North Sea. In southern England uplift of the Anglo-Brabant Landmass led to the input of coarse clastic sediments into the adjacent sedimentary basins. In one of these basins, the Wealden Basin, a landscape of braided rivers and mudflats developed. A significant rise in sea level took place towards the end of the Early Cretaceous and rifting of the Atlantic was temporarily abated such that the eastern margin of the young Atlantic became essentially passive (Figure 11.8). This transgression brought marine conditions back to southern England and the Midlands. Mudrocks, glauconitic sandstones and some limestones indicative of a shallow sea environment were deposited over much of southern and eastern England, as well as in the North Sea. The North Sea high areas were flooded for the first time during this transgression.

By the Late Cretaceous, sea floor spreading had been initiated in the Atlantic region, with the opening of the Rockall trough, but the later part of the Cretaceous is in general a tectonically quiet period with sedimentary basins produced by simple downwarps (Map 13). To the south, the Tethys Ocean, which separated Africa from Europe and Asia, was closing as the components of Pangaea and particularly Gondwana, began to separate. A rise in eustatic sea level, the most dramatic in Earth history (Figure 10.8) gave rise to the flooding of a large part of north-west Europe, including Ireland and most of Britain. Over much of this area a planktonic ooze, rich in **coccoliths** (Figure 10.6), the calcareous remains of a marine plant, was deposited to give **chalk**. The accumulation of chalk continued until the end of the Cretaceous in eastern England. Tectonically, the compressive regime produced to the south of Britain by the closure of the Tethys Ocean led ultimately to the inversion of these sedimentary basins and their uplift. This, together with a major regressive phase, led to the emergence and erosion of much of the younger chalk outcrop. In addition to this regressive event the Late Cretaceous is associated with a deterioration in climate which ultimately led to the onset of the Cenozoic 'Ice Age' (Figure 10.8). Finally, the end of the Cretaceous is also associated with a mass extinction (Figures 4.10 and 10.8), perhaps associated with a meterorite impact, evidence of which has been recorded from many sites across the world, but yet again there is no evidence for this event in the sedimentary record of onshore Britain, which was raised above sea level at this time.

The Palaeogene was dominated by sea floor spreading in the Atlantic, crustal thinning in the North Sea and by the closure of the Tethys Ocean which led to the Alpine Orogeny in southern Europe and Asia (Figure 11.8). Sea floor spreading in the Atlantic commenced in the Late Cretaceous, but large-scale activity did not commence until the Palaeogene. This resulted in widespread igneous activity either side of the diverging plate margin (Map 14). A suite of lavas, batholiths and dyke swarms were extruded and emplaced in north-west Britain, which led to the regional uplift of the ancient Caledonian Mountains of Scotland. The results of this volcanic activity are often referred to as the **Tertiary Volcanic Province**, which actually forms part of a much larger episode of volcanic activity associated with the rapid opening of the North Atlantic (Map 14). At the same time the North Sea was the site of rapid subsidence, such that marine conditions continued to dominate in this area during the Palaeogene. Subsidence of the North Sea was caused by cessation of thermal upwelling (i.e. molten rock) which had caused crustal extension in this area on and off during the Mesozoic. In fact, many would argue that the North Sea is perhaps best viewed as a rift complex which did not quite make it into a zone of plate divergence. If it had, Britain may by now have been located on the American side of the Atlantic. The uplift of northern Scotland during the Palaeogene provided a source of clastic debris for the North Sea Basin. In contrast to, and probably as a result of, the thermal uplift in north-west Britain, southern Britain was experiencing subsidence. In conjunction with this subsidence, minor sea level fluctuations occurred, which were probably the result of variation in the rate of sea floor spreading within the Atlantic today. The delicate interplay of subsidence and sea level variation in southern Britain resulted in a cyclic transgressive–regressive facies (Box 5.2). The transgressive phases within this cycle are recorded by marine sands and muds, while the regressive episodes are indicated by fluvial sands. The maximum marine advance in southern Britain took place in the Eocene, and deposited the marine muds of the London Basin, the London Clay.

During the Late Palaeogene the effects of the Alpine Orogeny, caused by the closure of the Tethys Ocean, were felt in Britain. The Alpine Front is far to the south of Britain and comprises the Alpine, Carpathian and Himalayan mountain belts. Large-scale nappe tectonics developed throughout the Alpine Front with maximum compression in the Late Palaeogene and Early Neogene. The effects in Britain were restricted to basin inversion and folding of Palaeogene sediments and the gentle deformation of the Cretaceous basins in the Neogene. This was coincident with vigorous sea floor spreading in the Atlantic which caused widespread uplift in the British Isles, and this trend was reinforced by the compressional regime of the Alpine Orogeny.

In the Neogene there was a small marine transgression which resulted in the deposition of marine shelly sands in East Anglia (Map 15). The marine molluscs of these 'Crag' deposits show that the climate was cooling rapidly towards the Cenozoic 'Ice Age' (Figure 10.8). The fossil molluscs present give valuable information on the climate as they are similar to present-day molluscs which range from warm-temperate to subarctic species. The cooling trend experienced in the Neogene culminated at the start of the Quaternary with the development of mid-latitude ice sheets in northern Europe and North America (Box 10.3).

Much of the British Isles was glaciated several times during the Quaternary as the mid-latitude ice sheets grew and decayed in the Northern Hemisphere. Periods of glacial expansion are known as **glacials** while those in which the mid-latitude ice sheets are absent are known as **interglacials** (Figure 11.9). We are currently within an interglacial and mid-latitude ice sheets may again accumulate in the Northern Hemisphere some time within the next few thousand years as we enter the next glacial. The oscillation between glacial and interglacials during the Quaternary is controlled for the most part by Milankovitch radiation variations (Box 10.1). A full record of the global oscillation between glacial and interglacial conditions is recorded in the oxygen isotope record, but unfortunately the record of these events in Britain is fragmentary since recent glacial advances erode and destroy the evidence of earlier episodes. Consequently, we know a lot about recent events but very little about the Early Quaternary history of Britain. The most recent glacial is known as the **Devensian** in Britain and the extent of the British ice sheet during this period is shown in Map 16.

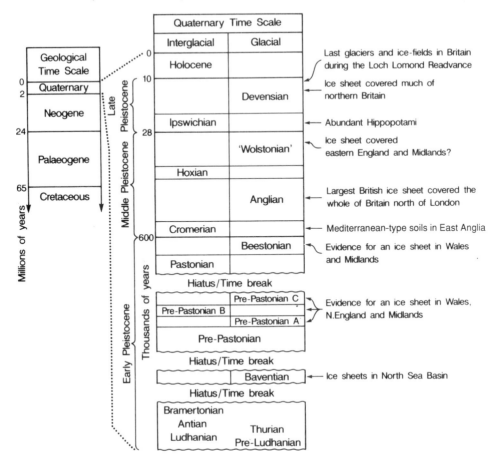

Figure 11.9 *The stratigraphy of the Quaternary 'Ice Age' deposits. [Based on information in: Boulton (1992). In: Duff & Smith (Eds)* Geology of England and Wales, *Geological Society of London, pp. 413–444]*

KEY EVENTS: THE LONG QUIET PERIOD

Plate tectonics: With the development of the large supercontinent at the start of the Permian, the tectonic regime switched from one of compression to one of extension which ultimately caused the break-up and dispersal of the supercontinent. Britain was located to the east of one of the major rift axes, and experienced periods of extensional tectonics and volcanism associated with the break-up of Pangaea.

Climate: During the early part of this interval the Earth experienced some of the warmest conditions within its history, before a gradual cooling took place in the Cenozoic which finally resulted in the establishment of an 'Ice-house' state. The period of global warmth is recorded in Britain by the deposition of extensive desert deposits, including redbeds. Within the British Isles, the return to 'Ice-house' conditions resulted in the growth and decay of a succession of ice sheets.

Sea level: With the construction of Pangaea, eustatic sea level reached an all-time low. With the onset of extensional tectonics, however, eustatic sea level began to rise steadily throughout the latter part of the Mesozoic and Early Cenozoic, reaching its high stand in the Late Cretaceous. This trend was reversed as polar ice sheets began to develop, as the Earth returned to an 'Ice-house' state.

Biosphere: During this period there were several major crises within the biosphere, none of which are adequately recorded in the British Isles due to the absence of marine conditions at the appropriate times. The high sea levels and global warmth during part of this interval resulted in the widespread deposition of chalk, which is composed of microscopic components of marine plants.

11.6 SUMMARY OF KEY POINTS

It is apparent from the above narrative that the stratigraphical record within the British Isles appears to reflect the global rhythm in plate tectonics, climate, sea level and the biosphere identified in Chapter 10. The rocks of the British Isles can be explained in terms of the subtle interplay of these four variables. The geological record in Britain is dominated by the assembly of Pangaea in the Palaeozoic and with the opening of the North Atlantic in the Late Mesozoic and Cenozoic. The oscillation between 'Ice-house' and 'Green-house' states can also be recognised, although it is modified by Britain's northward drift through the equatorial belts during the Late Palaeozoic. This is particularly true of the Carboniferous during which the global climate was generally cold, but Britain—located in the tropics at that time—experienced tropical conditions and shows little direct evidence of this cold phase. The pattern of eustatic sea level fluctuations has also had a profound effect on the nature of Britain's stratigraphical record, particularly during the long quiet period of the Mesozoic. During this period a succession of marine transgressions and regressions resulted in the deposition of a complex succession of different facies, which have contributed significantly to the diverse range of rocks present in Britain today. Some of the major events of biosphere development are, however, less well documented in Britain's stratigraphical record. For example, there is little or no record of many of the later mass extinction events of the Phanerozoic, primarily because Britain was experiencing uplift and erosion during many of these events and consequently few fossils were preserved. However, Britain does contain an excellent record of many of the Early Palaeozoic events. In general, Britain's geological record is remarkable for its completeness and for the detail with which the pulse or rhythm of global events has been recorded.

11.7 PALAEOGEOGRAPHICAL MAPS OF THE BRITISH ISLES

The following palaeogeographical maps show the changing pattern of land, sea and sedimentary environments within the British Isles since the Devonian. In many ways these maps simply represent a 'cartoon-strip' of the development of the British Isles.

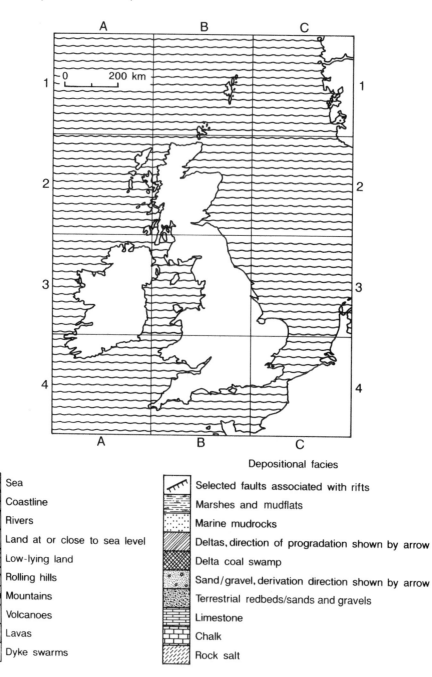

Depositional facies

Sea	
Coastline	
Rivers	
Land at or close to sea level	
Low-lying land	
Rolling hills	
Mountains	
Volcanoes	
Lavas	
Dyke swarms	

Selected faults associated with rifts	
Marshes and mudflats	
Marine mudrocks	
Deltas, direction of progradation shown by arrow	
Delta coal swamp	
Sand/gravel, derivation direction shown by arrow	
Terrestrial redbeds/sands and gravels	
Limestone	
Chalk	
Rock salt	

Map 1 *Key to palaeogeographical maps*

Late Devonian - *c.*370 million years

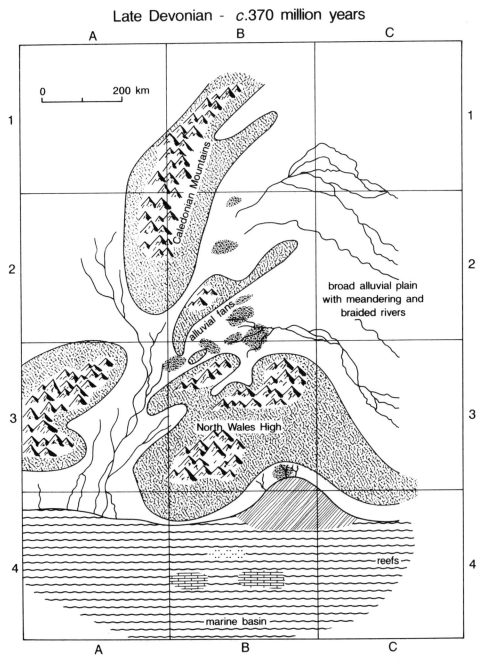

Map 2

Maps 2–15 are all based on information in Ziegler (1990) Geological Atlas of Western and Central Europe, *Geological Society Publishing House; and Cope et al. (Eds) (1992)* Atlas of Palaeogeography and Lithofacies, *Geological Society Memoir No. 13*

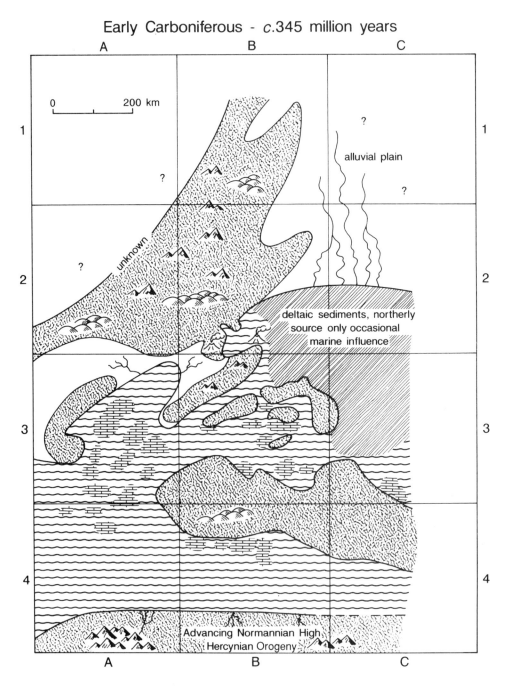

Early Carboniferous - *c.*345 million years

Map 3

Mid-Carboniferous - *c*.315 million years

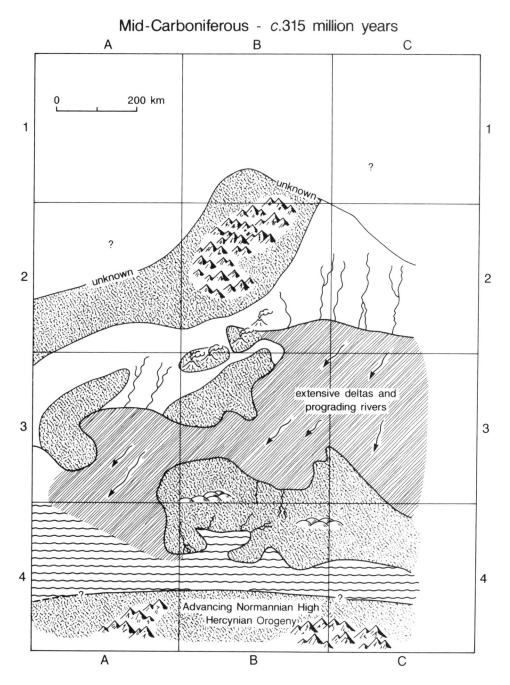

Map 4

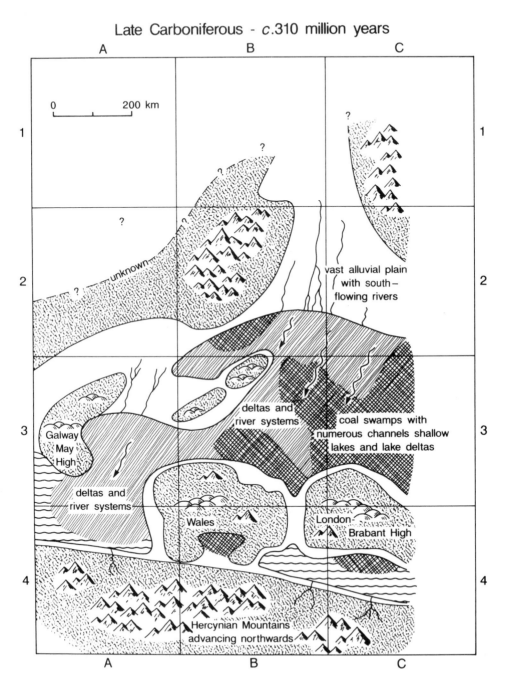

Late Carboniferous - *c*.310 million years

Map 5

Early Permian - *c*.260 million years

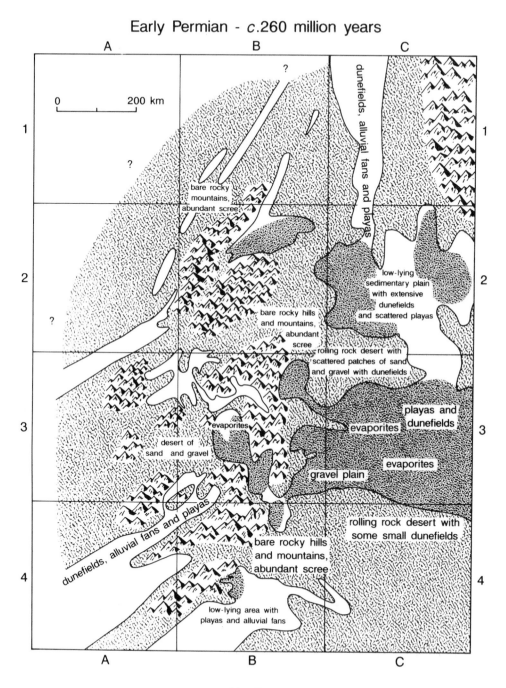

A B C

0   200 km

?

?

?

dunefields, alluvial fans and playas

bare rocky
mountains,
abundant scree

low-lying
sedimentary plain
with extensive
dunefields
and scattered playas

bare rocky hills
and mountains,
abundant
scree

rolling rock desert with
scattered patches of sand
and gravel with dunefields

playas and
dunefields

evaporites

evaporites

desert of
sand and gravel

evaporites

gravel plain

rolling rock desert with
some small dunefields

dunefields, alluvial fans and playas

bare rocky hills
and mountains,
abundant scree

low-lying area with
playas and alluvial fans

1 2 3 4

Map 6

Late Permian - *c.*255 million years

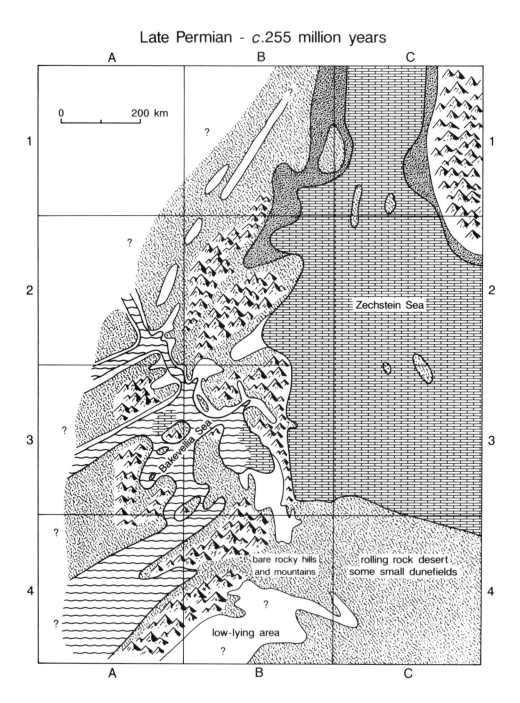

Map 7

Mid-Triassic - *c*.235 million years

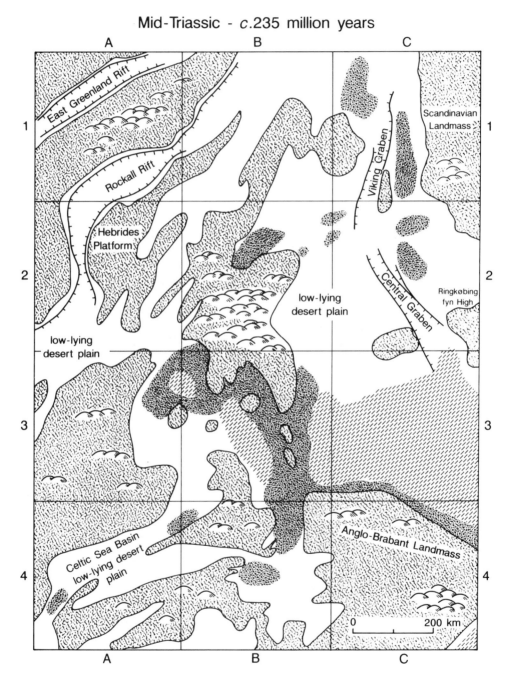

Map 8

Early Jurassic - *c*.190 million years

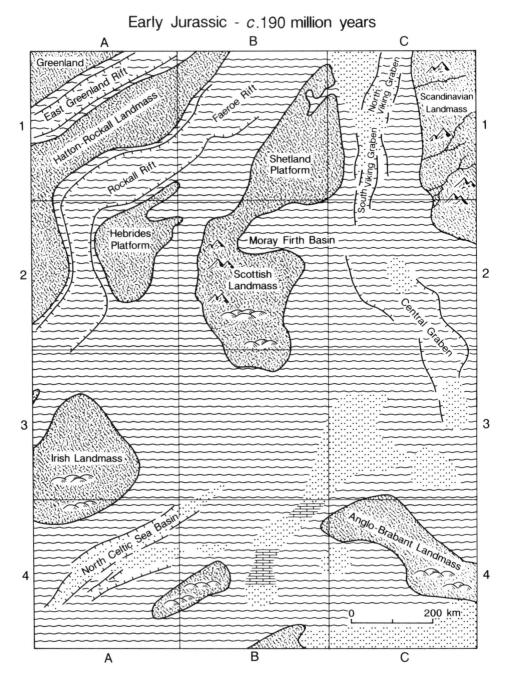

Map 9

Mid-Jurassic - *c*.175 million years

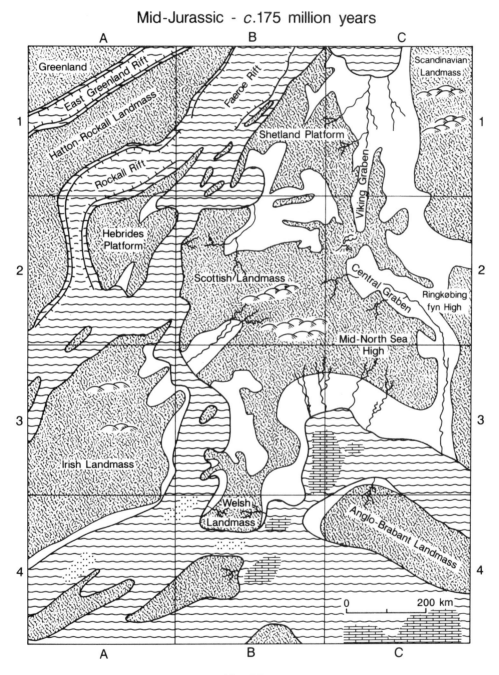

Map 10

Late Jurassic - *c.*150 million years

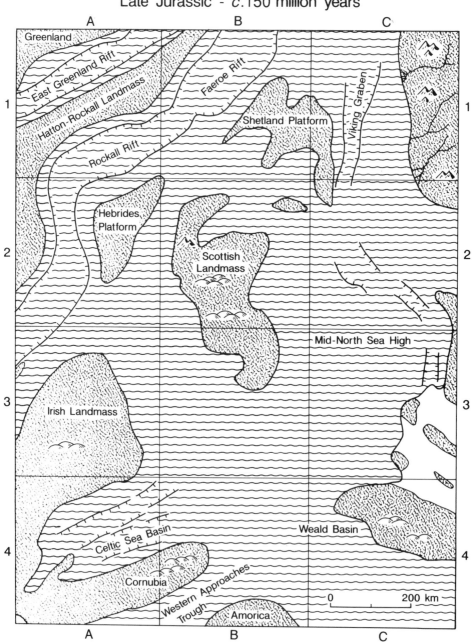

Map 11

Early Cretaceous - *c*.130 million years

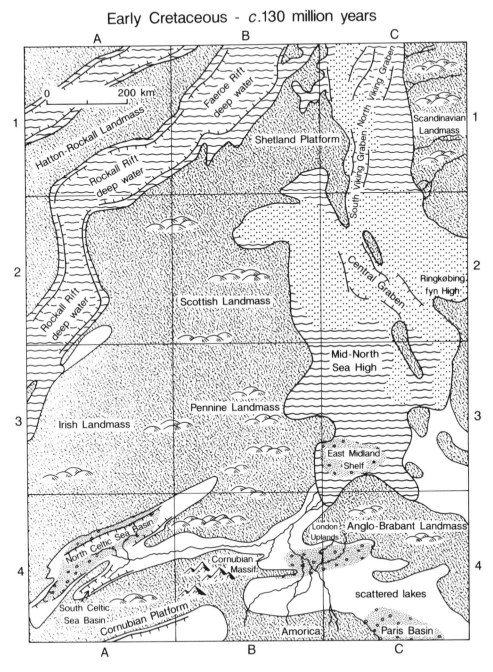

A   B   C

Hatton-Rockall Landmass

Faeroe Rift deep water

Shetland Platform

South Viking Graben North Viking Graben

Scandinavian Landmass

1

Rockall Rift deep water

Rockall Rift deep water

2

Scottish Landmass

Central Graben

Ringkøbing-fyn High

Mid-North Sea High

2

3

Irish Landmass

Pennine Landmass

East Midland Shelf

3

North Celtic Sea Basin

Cornubian Massif

London Uplands

Anglo-Brabant Landmass

scattered lakes

4

South Celtic Sea Basin

Cornubian Platform

Amorica

Paris Basin

4

A   B   C

0 200 km

Map 12

Late Cretaceous - *c*.75 million years

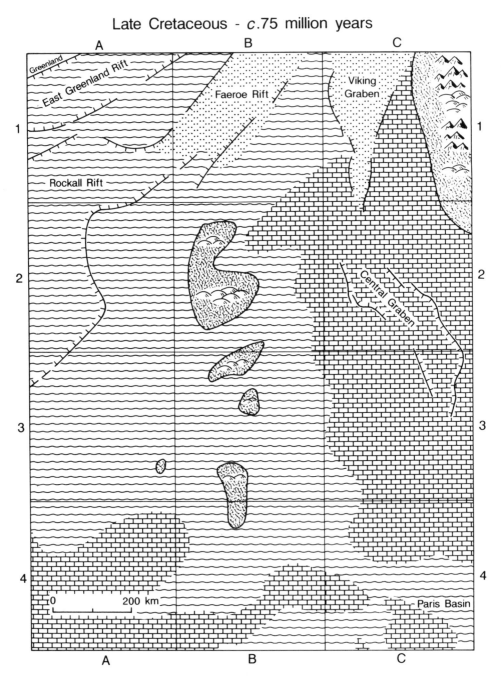

Map 13

Early Cenozoic (Palaeogene) - *c*.60 million years

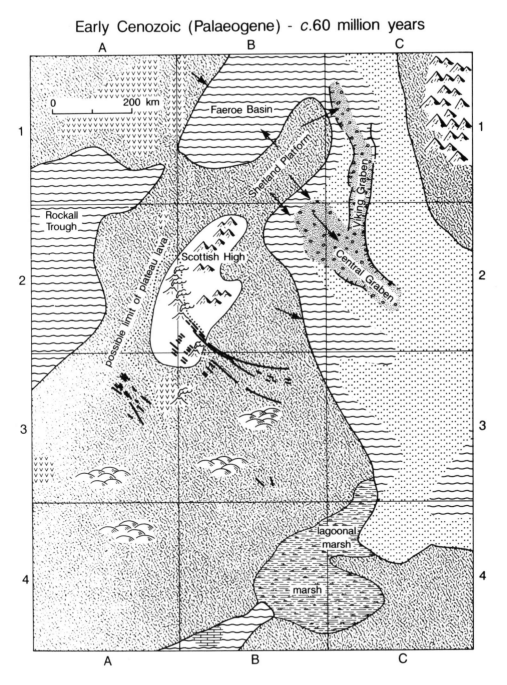

Map 14

Late Cenozoic (Neogene) - *c*.15 million years

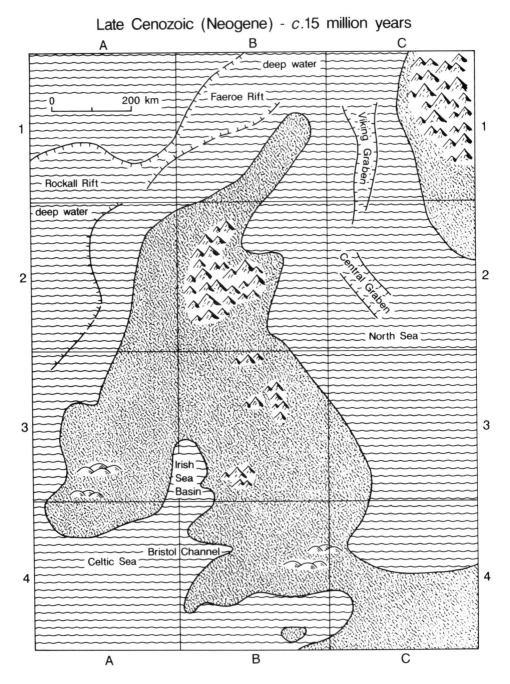

Map 15

Late Cenozoic (Quaternary) - *c.*20 000 years

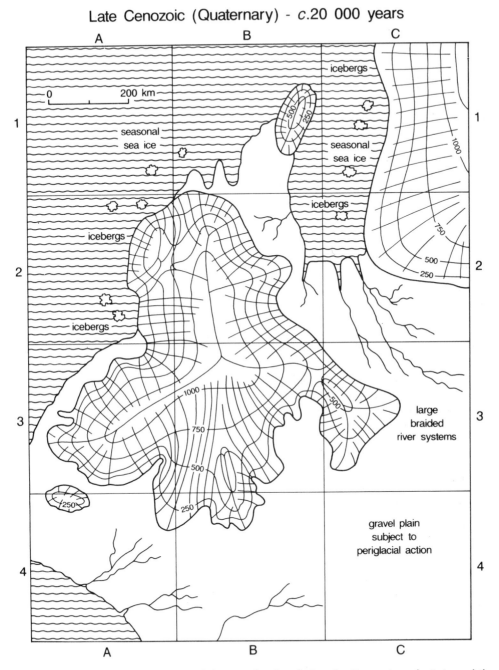

Map 16 *Palaeogeographical map of the British Isles during the Devensian glaciation of the Cenozoic 'Ice Age'. [Based on information in: Boulton (1991) In: Craig (Ed.) The Geology of Scotland, The Geological Society, Fig. 14.4, p. 406]. Principal flow lines and ice surface contours (in metres) are shown*

11.8 SUGGESTED READING

Much of the information covered in this chapter is dealt with in more detail by Anderton *et al.* (1979) and Rayner (1981). More condensed accounts are provided by Owen (1976) and Dunning *et al.* (1978) although these are now a little dated. Cope *et al.* (1992) contains a series of detailed palaeogeographical maps for the British Isles, which are accompanied by excellent summaries and both provide a vital source of information. This information can be supplemented by reference to Ziegler (1990). The vast literature on the stratigraphy of the British Isles is reviewed in Craig (1991) and in Duff and Smith (1992), both of which contain excellent bibliographies, which provide a means of access to the wealth of literature on the subject. A detailed consideration of the onshore sedimentary basins of England and Wales is given by Whittaker (1985). Glennie (1990) provides an important North Sea perspective to British stratigraphy. An up-to-date account of the tectonic evolution of the British Isles in the context of northern Europe is provided by Berthelsen (1992).

Anderton, R., Bridges, P. H., Leeder, M. R. & Sellwood, B. W. 1979. *A Dynamic Stratigraphy of the British Isles.* Allen & Unwin, London.

Berthelsen, A. 1992. Mobile Europe. In: Blundell, D., Freeman, R. & Mueller, S. (Eds) *A Continent Revealed: The European Geotraverse.* Cambridge University Press, Cambridge, 11–32.

Cope, J. C. W., Ingham, J. K. & Rawson, P. F. 1992. *Atlas of Palaeogeography and Lithofacies.* Geological Society Memoir No. 13.

Craig, G. Y. (Ed.) 1991. *Geology of Scotland.* (Third Edition) Geological Society, London.

Duff, P. McL. D. & Smith, A. J. (Eds) 1992. *Geology of England and Wales.* Geological Society, London.

Dunning, F. W., Mercer, I. F., Owen, M. P., Roberts, R. H. & Lambert, J. L. M. 1978. *Britain before Man.* HMSO, London.

Glennie, K. W. 1990. *Introduction to the Petroleum Geology of the North Sea.* (Third Edition) Blackwell, Oxford.

Owen, T. R. 1976. *The Geological Evolution of the British Isles.* Pergamon Press, Oxford.

Rayner, D. H. 1981. *The Stratigraphy of the British Isles.* Cambridge University Press, Cambridge.

Whittaker, A. (Ed.) 1985. *Atlas of Onshore Sedimentary Basins in England and Wales: Post-Carboniferous Tectonics and Stratigraphy.* Blackie, Glasgow.

Ziegler, P.A. 1990. *Geological Atlas of Western and Central Europe.* (Second Edition) Geological Society, London.

12
Summary of Part II: The Pattern of Earth History

The traditional approach to stratigraphy is to build up a picture of Earth history layer-by-layer. Only when all the layers are in place can one pick out the global pattern and the broad controls on the development of the stratigraphical record. In the second part of this book we have tried to take an alternative approach. First, we have examined the global pattern of temporal change within four key variables which control the nature of the Earth's physical environment—plate tectonics, climate, sea level and the biosphere. Second, we have tried to show how this pattern of global change can be used as a framework with which to examine, at the broadest scale, the geological development of the British Isles.

In Chapter 10, we explored the global rhythm of change within each of the four key variables which control the physical environment. The changing pattern of continents and plates can be viewed as a large cycle—the supercontinental cycle—in which the Earth's continents have been first dispersed and then assembled into a large supercontinent. The Phanerozoic records such a cycle, and there may have been at least one more in the Precambrian. The Earth's climate has oscillated throughout the Phanerozoic between periods of global refrigeration ('Ice-house' states) and periods of global warmth ('Green-house' states). Sea level has also fluctuated during this time with rises and falls responding to both tectonic and climatic patterns of change. Finally, the biosphere has developed through time as life on Earth has evolved and diversified, periodically suffering extinctions.

In Chapter 11, we examined the geological development of one small portion of the Earth's crust with reference to this framework or pattern of global events. Our chosen example was the British Isles, although we could equally well have selected almost any continental area (e.g. North America, Australia, or even Antarctica). From this examination, it is clear that in broad outline the geological history of the British Isles does appear to reflect the rhythm of global changes. The pattern is, however, modified by changes in Britain's latitude which overprint the global pattern, but even this serves to illustrate the overall importance of plate tectonics in driving Britain northwards through time. Similarly, not all the global variations in the biosphere are recorded, such as mass extinctions in the marine realm, because of the absence of marine environments in Britain at critical points. The global pattern may also be obscured when one looks at the detail within the stratigraphical record, where regional and local controls may be of more significance. Here the global variation will be present, but obscured by small-scale or local uplift, sea level changes and climate.

In Part II we have provided a framework of global change which you will be able, with the tools presented in Part I, to apply to any part of the Earth's surface in order to understand its geological history.

Index